Unfulfilled Promise

Collective Bargaining in California Agriculture

Philip L. Martin, Suzanne Vaupel, and Daniel L. Egan

Routledge
Taylor & Francis Group

LONDON AND NEW YORK

First published 1988 by Westview Press, Inc.

Published 2018 by Routledge
52 Vanderbilt Avenue, New York, NY 10017
2 Park Square, Milton Park, Abingdon, Oxon OX14 4RN

Routledge is an imprint of the Taylor & Francis Group, an informa business

Copyright © 1988 Taylor & Francis

Library of Congress Cataloging-in-Publication Data
Martin. Philip L.
  Collective bargaining in California agriculture.
  (Westview special studies in agriculture science and
policy)
  Includes index.
  1. Collective bargaining—Agriculture—California.
2. Agricultural laborers—California. 3. Industrial
relations—California. I. Egan, Daniel L. II. Vaupel,
Suzanne. III. Title. IV. Series.
HD6976.A292U66 1988   331.89′043′09794   88-5632

ISBN 13: 978-0-367-21278-0 (hbk)
ISBN 13: 978-0-367-21559-0 (pbk)

Unfulfilled Promise

Dedicated to Jessica, Patrick, Tom and Mary

Contents

Tables and Figures

Tables

Figures

Foreword

The "plight of the California farmworker" has been the main theme of over 100 years of government reports, scholarly writings, and popular literature. Farmworkers were excluded from most of the 1930s legislation which regulated wages and working conditions and recognized that workplace disputes could best be settled by collective bargaining. Scholars and advocates argued that the nation's several million farmworkers were entitled to the same protections available to nonfarm workers, but farmers successfully argued that agriculture was unique and that labor laws were neither necessary nor workable on farms.

California employs about one-fourth of the nation's hired farmworkers. In many instances, crews of several hundred workers are brought together in open-air factories to cultivate and harvest fruits and vegetables. For over a century, there was no governmental framework to resolve workplace disputes.

In 1975, California became the first major agricultural state to establish a framework "to bring peace and stability to agricultural labor relations." The Agricultural Labor Relations Act granted farmworkers the right to organize and to bargain collectively with farm employers and established an administrative agency to enforce these workers' rights.

This book puts this recent farmworker experience in perspective. The book explains the importance of farmworkers in California agriculture, reviews the history of farm labor and farmworker unions, and then examines the ALRA in operation.

The ALRA and its administrative agency, the ALRB, have been controversial experiments to empower the powerless. This book helps to explain why the law and agency which Cesar Chavez and the United Farm Workers fought for in the 1970s is viewed as the farmworkers' enemy in the 1980s. Perhaps it was inevitable that essentially a one union-one industry labor law would become embroiled in politics, but the political swings in the interpretation and administration of the ALRA serve as a sober reminder of the polarized nature of the farmworker debate in the United States.

This book is a well-written examination of the origins and operation of California's Agricultural Labor Relations Act. It also provides useful insights into the problems involved in bringing collective bargaining rights to farmworkers. Indeed, California provides what is probably a best case example, but we can learn national lessons from it. This book is therefore a highly recommended introduction to the persisting farm labor controversy.

Ray Marshall
University of Texas

Preface

California is the most important agriculture state in the United States, and farming is one of the state's largest industries. California agriculture differs from farming elsewhere because large growers bring together hundreds of seasonal workers in "factories in the fields" to cultivate and harvest almost half of the nation's fruits and vegetables. For a century, the proper relationship between these farmworkers and their employers has been debated.

Forty years after the National Labor Relations Act excluded farmworkers "for administrative reasons" from the framework established to regulate the interaction of workers, employers, and unions, California enacted an Agricultural Labor Relations Act (ALRA) "to bring certainty and a sense of fair play" to labor relations in agriculture. The ALRA added the missing ingredient to farm labor relations by granting farmworkers the right to organize and by promising farm employers labor stability.

The ALRA has not fulfilled its promise. Farm labor relations are still considered "immature." ALRA and its administrative agency, the Agricultural Labor Relations Board (ALRB) have been controversial, as both growers and unions alternately charged that the ALRA and ALRB are "biased" against them. A decade after California acted to bring peace and tranquility to the fields, the ALRB has not yet taken root and flourished as the preferred agency to help resolve the inevitable disputes between workers, unions, and employers.

The purpose of this book is to put the California farmworker union experience in perspective. This goal is accomplished by explaining the importance of farmworkers in California agriculture, outlining the history of farm labor and farmworker unions, and then examining the ALRA and ALRB.

This book explains why a labor relations law was considered necessary, the provisions of the ALRA, and how the ALRB implements the ALRA. Readers who understand the historical, economic, and legal background of farmworker unions can make an independent assessment of the charges which still surround labor relations in California agriculture.

Philip L. Martin
Suzanne Vaupel
Daniel L. Egan

Acknowledgments

This book evolved from five years of teaching UCD students about collective bargaining in California agriculture. We owe a debt to the students who struggled with an ever-expanded and revised manuscript.

One student, Stephanie Luce, became a valuable research assistant for this project by preparing data and checking the legal citations. Carole Nuckton edited the manuscript, and Stephen Sosnick and John Mamer provided helpful reviews. Kelly Cochran prepared the camera ready copy of this book.

A project which is refined over several years owes much to its visionary supporters. Kirke Wilson of the Rosenberg Foundation has been a consistent supporter of our farm labor research, for which we are grateful. Support for particular phases of the project was received from the UCD Teaching Resources Center, the Giannini Foundation, and the Agricultural Issues Center. We are also grateful to the UCD Agricultural History Center and the Institute of Governmental Affairs for their continuing support.

Philip L. Martin
Suzanne Vaupel
Daniel L. Egan

Chapter 1

Farmworkers in California Agriculture

Agriculture is the heart of the food system—the largest industry in the United States and in California. The U.S. food system that manufactures and distributes farm inputs, produces food and fiber on farms, and packs and processes farm products for consumers generates about 18 percent of the gross national product and employs 18 percent of the work force—2 percent work to supply agriculture with equipment, seeds and fertilizers, and other inputs; another 2 percent consists of farmers and farmworkers; and 14 percent of all U.S. workers are employed by food processors, packers, retailers, and distributors. This food system is considered a paradigm for emulation because it provides such a variety of high quality and low-cost foods to consumers. The average American family spends about 16 percent of its disposable income on food, substantially less than the 20 to 25 percent common in Europe.[1]

Farm labor has been a perennial issue since the dawn of American agriculture. American farming eventually evolved into three distinct labor systems. The family farm of the Northeast and Midwest was usually a diversified and self-sufficient livestock and field crop enterprise that produced primarily for the family with family labor. Commercial cotton and tobacco farms in the Southeast were dependent first on slaves and later on sharecroppers and tenants to produce for distant markets. As grain production moved further west, farmers exchanged labor at harvest time to produce grains for distant markets, but this exchange labor was supplemented by migrant farmworkers who followed the harvest from farm to farm. California agriculture, by contrast, has been dependent for over a century on migratory farmworkers who are employed only seasonally.

The share of the U.S. work force engaged in farming began to fall in the 1920s. Farmworkers and farmers were pushed out of agriculture by labor-saving equipment and low farm prices and pulled into nonfarm jobs by higher wages since the 1940s. After sharp declines in the number of farmers and farmworkers in the

1950s, total farm employment began to stabilize in the 1970s. The number of farmers continued to decline, but this decline was offset by an increase in the number of hired farmworkers. Hired farmworkers, who had done about 25 percent of U.S. farm work from 1910 to 1970, did 35 percent of the nation's farm work by 1980.[2]

Within U.S. agriculture, farm labor markets differ by commodity and region. Three major commodity groups divided the $10 billion farm wage bill in 1985 about equally, but wages were a much higher fraction of fruit, vegetable and horticultural (FVH) specialty sales ($21 billion), than of field crop sales ($50 billion) or livestock sales ($69 billion). Farmers in each of these farm sectors employ domestic and foreign seasonal workers, but the major employer of seasonal workers is the FVH subsector.

This chapter examines farm labor issues in California agriculture. The first section contrasts the structure of California agriculture with farming elsewhere. The second and third sections review the demand for and supply of farmworkers, and the fourth section analyzes the operation of the farm labor market and examines the effects of collective bargaining on farmworker wages. The fifth section concludes that the farm labor market is being "balkanized" by being tugged in opposite directions: technology and collective bargaining are raising wages and integrating some farm and nonfarm labor markets, but illegal immigration is simultaneously reducing wages and further isolating other farm labor markets. Thus, California's farm labor market is being split into distinct commodity and regional segments, making local factors the most important determinants of farm wages and working conditions.

A. CALIFORNIA AGRICULTURE

California's 82,500 farms produce over 250 commodities that range from milk and cotton to grapes and artichokes. Agriculture is a major industry in California, employing over 1 million farmers and workers sometime during each year to produce farm products worth $14 billion. California agriculture has several distinguishing features, including multiple cropping on large acreages of high-value, irrigated cropland. The most salient feature for labor market analysis is the importance of commercial fruit and vegetable farming; in 1985, FVH sales of $8 billion were 80 percent of California crop receipts. Some of the FVH commodities such as processing tomatoes, potatoes, and most nuts are harvested mechanically, but the others require large numbers of year-round and seasonal workers for pruning, cultivating, irrigating and harvesting.[3]

Fruit, vegetable, and horticultural specialty farms, regardless of their size, tend to rely on hired workers. Farmers do 65 percent of the nation's farm work, but in California hired workers do over 75 percent of the state's farm work and often all of the farm work on commercial fruit and vegetable farms. Instead of farm owners hiring farmworkers directly, employers are often hired managers, field supervisors, or independent farm labor contractors who negotiate a price for recruiting and supervising a crew of workers to perform a specific farm task such as harvesting. The importance of hired workers and absentee landowners who depend on managers and labor contractors helps to explain why California agriculture reverses a familiar phrase: farming in California is a business and not a way of life.

The changing structure of farming in California parallels national trends: middle-sized farms are being pushed either toward the small or large extremes of the farm-size spectrum. Most of California's 82,500 farms are small and part-time operations: almost half sold less than $10,000 worth of farm products in 1982, and less than half of all "farmers" reported that farming was their principal occupation. A handful of very large farms produce most of California's farm products: 8,800 farms had annual sales of $250,000 or more in 1982, and these farms accounted for almost 90 percent of California farm sales.

California agriculture resembles and differs from agriculture in other states. California's climatic diversity means that the state's farming ranges from mountain ranching to cash grain farming to labor-intensive fruits and vegetables. There are both large and small farm employers: half of California's farms employ no hired workers, and many employ just one or two, but there are also a handful of large labor-intensive farms. However, the production of many fruits and vegetables is concentrated on just a handful of farms, so that most farm labor issues involve fewer than 10 percent of the state's farms. For example, the 10 largest lettuce growers produce about 50 percent of the lettuce, and the largest producers have $20 to $25 million annual wage bills.[4] Since California produces the lion's share of many fruits and vegetables, including three-fourths of the nation's lettuce and strawberries and one-fourth of the nursery products, concentration of fruit and vegetable production in California means that a large share of the nation's labor-intensive commodities are produced on a handful of seasonal "factories in the field" with hired managers and hired workers.

Fruit and vegetable production in California is "industrial agriculture" in the sense that large family and nonfamily corporations borrow capital, buy or rent land, hire production managers and workers, and establish subsidiaries to handle the packing and transportation of a substantial fraction of the crop.

B. FARMWORKER EMPLOYMENT

California's farm labor market reflects the peculiar nature of fruit and vegetable production. On most large farms, a handful of year-round or permanent employees are joined by a fluctuating number of seasonal workers who handle particular jobs: pruning trees and vines in winter, thinning, hoeing, and irrigating during the spring and summer, and harvesting in the fall. A few labor-intensive commodities do not fit this peak and trough mold. Nurseries and greenhouses offer year-round work at one location, and some large vegetable growers own or rent land in several areas and offer almost year-round work to employees who move with the harvest. Commodities such as citrus, strawberries, and broccoli can offer six or seven months employment in one location, but most seasonal farmworkers are employed on one farm for only two to six weeks.

The demand for labor in California agriculture rises and falls throughout the year. In the mid-1980s, seasonal farmworker employment was estimated to be 78,000—its lowest point—in March and its highest—182,000—in September, a peak-trough ratio of 2.5.[5] Peak-trough ratios vary considerably within the state, e.g., 10 in Imperial County, six in Fresno County, and three in Monterey County, with the highest ratios indicating more seasonality. Employment fluctuations mean that three different labor market indicators—average employment, total work force, and hours worked—are needed to obtain a full picture of farmworker employment.

The farm labor market indicator most often discussed is average employment, the average number of persons employed during a particular time period such as one day or one week. In California, estimates of average employment are made for one week during each month for farmers, regular workers employed by one employer at least six months, and seasonal workers. During 1985, average employment was 275,500—an average 60,700 farmers, 93,900 regular workers, and 120,600 seasonal workers were employed during the 12 weekly surveys.[6] Unemployment insurance data indicate higher average employment: an average 236,000 hired workers were employed on crop and livestock farms in 1985.

The total work force indicates how many persons filled these "year-round equivalent jobs" available in California agriculture. If all farm jobs were year-round and there were no worker turnover, the total work force would equal average employment. Since farm jobs begin and end, and farmworkers enter and leave the farm work force, the total farm work force exceeds average employment. In 1985, about 600,000 workers filled the average 236,000 year-round equivalent jobs available on crop and livestock farms, a ratio of 2.5 workers for each year-long equivalent job.[7] A year-long equivalent seasonal job is created by adding-up e.g., one seasonal worker working six months, another three, another two, and another one month. It is important to realize that because most seasonal farm jobs occur in the summer and fall months, the number of workers who are employed in seasonal agriculture is three or four times the number that would be needed if all farm jobs were year-round.

This seasonal demand for farmworkers—the need to hire 4 or 5 workers in September for every worker employed in February—means that a massive mobilization of farmworkers occurs every summer and fall. Farmers worry that not enough workers will be available to harvest their crops in time to prevent weather and other damage, while workers fear that too many workers will show up and prevent them from earning enough to tide them over until next year's harvest.

Seasonality breeds uncertainty for farmers and workers; and the farm labor market has not developed mechanisms to minimize uncertainty. Most farmers prefer to maximize the pool of potential seasonal workers so that enough workers will be available to fill all seasonal jobs, and few workers are willing to assure that they will return on schedule next season.

The third farm employment indicator is the number of hours worked. Hours-of-work data are not available by state or by type of worker. California accounts for 75 percent of farm sales in the California-Oregon-Washington (Pacific) production area, however, and Pacific agriculture accounted for 10 percent of the nation's 6 billion farm work hours in 1985.[8] In these Pacific states, three-fourths of the farmer and farmworker hours were used to produce crops, with vegetables (160 million hours) and fruit and nuts (218 million hours) accounting for two-thirds of total hours worked in crops.

The major farm labor story in most industrial nations is the declining importance of farm work. In the United States, average farmworker employment fell 44 percent between 1950 and 1980; the total hired farm work force fell 37 percent; and total hours worked fell 70 percent.[9] California is an exception to this "rule" of a declining farm work force. Average employment was 218,000 in 1950 and 224,000 in 1980; the total hired farm work force rose from about 600,000 to about 1 million as agricultural service firms expanded;[10] and hours of crop work in the Pacific state declined only 55 percent since 1950. The number of farmworkers increased to replace the farmers who quit farming and to harvest crops on land that was double-cropped or brought into production when water became available in the 1950s and 1960s.

Average and total labor market indicators in California, however, may provide a misleading picture of employment stability because they exclude braceros—Mexicans who came to the United States to do farm work between 1942 and 1964. In 1957, the average employment of braceros in California peaked at 53,000 or 27 percent of average domestic farmworker employment. Braceros probably did a disproportionate share of California's farm work, perhaps as much as 40 to 45 percent, because they were experienced workers who came to the United States for one purpose: to do farm work.[11]

Hired workers remain crucial to California agriculture despite three decades of labor-saving mechanization because acreages and yields of labor-intensive fruits and vegetables have soared. During the 1950s, over 100,000 workers were employed to thin and harvest cotton, sugarbeets, and tomatoes; today, fewer than 20,000 workers are employed in these crops. Precision planting, labor-saving herbicides, and harvest mechanization also eliminated farm jobs, as did the reduction in the number of times a field was repicked and the switch to bulk or forklift handling of harvested commodities. However, these labor savings were offset by a doubling of fruit and vegetable production in California, the result of expanded acreages (e.g., grapes) and increased yields (e.g., strawberries). As production expanded and became more concentrated on commercial enterprises, more farmworkers were needed for irrigating and harvesting.

Production expansion dominated mechanization in the 1970s, increasing average farm employment in California about 6 percent during the decade.[12] Expansion may continue to offset mechanization as, for example, new strawberry and avocado jobs replace those lost by the adoption of machines to harvest wine grapes. The 1980s may also witness changes that alter the nature of farm jobs. These changes include automated drip irrigation systems; dwarf trees whose fruit can be harvested mechanically; increased imports of fruits and vegetables; and the field packing of vegetables and melons.

Almost all of California's harvested cropland is irrigated, and irrigation requires about one-sixth of the total hours of work in California agriculture. Irrigation traditionally requires workers to open and close furrow valves or move pipes, but the rising cost of water and energy and drainage problems are encouraging the spread of labor-saving drip irrigation systems that delivery a smaller quantity of water to each tree or vine. The result could be an increase in demand for labor to install the new irrigation systems, but a reduction in the labor needed to operate them.

Most of California's harvest workers are employed in fruits, and most fruits are picked by workers on ladders and dropped into bags that are draped over workers' shoulders. Plant scientists are developing dwarf trees whose fruit ripens simultaneously, and engineers are improving tree shaking and fruit catching devices. These scientific and technological changes could sharply increase the demand for labor to replant and prune smaller trees, but later reduce the demand for harvest labor.

California's fruit and vegetable agriculture expanded with affluence, population growth, and the movement of production from elsewhere in the United States to California. California exports labor-intensive commodities to distant markets within the United States and abroad, and it has faced relatively few competitive imports until recently. However, competition from producers in the East and Midwest in vegetables, from nations such as Spain and Israel in citrus, Italy and France in wine, Greece and Turkey in raisins, and Columbia in flowers pose formidable competitive challenges to California growers. Foreign producers, were helped by the strong dollar of the early 1980s, which made foreign products cheaper and U.S. exports more expensive.[13] If imports to the United States increase or exports decrease, production will eventually be reduced and farmworker employment will decline.

Farmworker employment may not decline as fast as increased mechanization and imports would suggest if harvesting and packaging improvements permit more vegetable and melon packaging to be done in the fields. Vegetable production on large acreages traditionally has separated field harvesting from (nonfarm) shed sorting and packaging, permitting the use of specialized workers and equipment in each operation. Field workers harvesting vegetables for piece-rate wages of say, $1, for each carton of lettuce packed, may earn $6 to $12 hourly, and packing shed wages may average $7 and $8 for workers who sort and pack the harvested commodity.

Field conveyor belts with packing platforms allow vegetable and melon growers to avoid packing shed costs by harvesting and packing in a single operation. These fieldworkers earn hourly wages of $4 to $6; hourly wages are paid because the employer controls the speed of the conveyor belt through the field and thus the pace of work. Many of these farmworkers are women and older men who are not suited to the rapid pace of piece-rate work which enables some young workers to earn $100 daily. Unions are apprehensive about field packing: it increases the number of "farmworkers" but it may also reduce the dominance of the large grower-packers whose workers are easiest to organize. Field packing might also renew conflict between farmworker unions if packing shed unions try to follow their members' jobs into the fields.

C. FARMWORKER CHARACTERISTICS

Farmers do about one-fourth of California's farm work; most (93 percent) are White. Most hired workers are Mexicans or Mexican-Americans, although Whites, Filipinos, Indochinese, Indians, Pakistanis, and Arabs also do farm work in California.[14] Generally, the Whites fill professional and clerical farmworker jobs in commercial agriculture, operate equipment on field crop farms, and are employed year-round in the livestock sector. Mexicans and Mexican-Americans are field foremen and workers, irrigators, and equipment operators, although some Mexicans also find employment on livestock farms. The other ethnic groups tend to specialize by commodity and region: Filipinos are concentrated in San Joaquin grapes, and the Pakistanis in northern California fruit and nut pruning and harvesting.

Many of the people employed in the "agriculture" industry do not have an agricultural occupation or job title, e.g., the agricultural industry employs managers, clerks, and workers with other job titles. In mid-1987, about two-thirds of the persons who were receiving unemployment insurance payments because they had worked for agricultural employers also had agricultural occupations such as farmworker, farm supervisor, or nursery worker. This difference between the occupation "farmworkers" and persons employed in the "agriculture" industry raises an important caveat about farmworkers: some people who satisfy the farmworker stereotype are not farmworkers, while other people who do not satisfy the stereotype are indeed farmworkers.

Physical effort in farm work ranges from relatively easy tasks such as operating equipment and sorting fruits and vegetables, to light hand tasks such as hoeing and irrigating, to heavy hand tasks such as manually harvesting fruit from ladders. Although all kinds of farmworkers can be found doing each of these tasks, there is a definite age and sex taxonomy. Younger workers dominate the heavy-harvesting tasks because hand-harvesting involves stooping, climbing, or carrying efforts that often "use-up" a worker's back in 10 to 15 years. Young men dominate the hand-harvesting of citrus and tree fruits, melons, and piece-rate vegetables, while older men and women often harvest strawberries, carrots, and other vegetables that are field packed.

Men do much of the irrigation work, while older men and women do most of the thinning and hoeing work. Older men and women fill many of the sorting and packing jobs in citrus, tree fruit, and vegetable packing sheds, and women also do much of the field sorting of mechanically harvested tomatoes. Men of all ages operate farm equipment.

Among the Mexican-heritage workers who fill farm jobs some are citizens, some legal immigrants (greencarders), and others are illegal aliens or undocumented workers. A 1983 survey of California's fieldworkers found that approximately 71 percent were Mexican born and that 22 percent of this Mexican subgroup were U.S. citizens, 55 percent were legal immigrants,[15] and 20 percent were illegal aliens. It appears that women fieldworkers are more likely to be legal workers than men are and that the most legal work forces tend to be in the high-wage Salinas and Imperial Valleys.

One important impact of the "Mexicanization" of the farm labor market is that Spanish has become the language of the fields. Growers look for workers in Mexico or the Mexican-American barrios of U.S. cities and towns; growers promote to supervisor workers who are bilingual; and most union organizing and other worker activity is in Spanish. Spanish is an important secondary language in California, but it is the most important language in the fields, and it helps to isolate unemployed White and Black workers from the farm labor market.

D. THE FARM LABOR MARKET

California's farm labor market matches over half a million seasonal workers with seasonal jobs each year. For many farmers, an entire year's farm income depends on harvesting a perishable commodity during a critical window of a few weeks. Farmworkers, who may find farm work for only 10 to 20 weeks each year, scramble to maximize their earnings during these harvest periods by finding the highest hourly or piece-rate wages. There is an inherent conflict-of-interest between farmers and workers. Farmers prefer a surplus of workers to minimize crop losses and wage costs, and workers prefer labor shortages to maximize wages and work time.[16] Farmers complain of labor shortages, yet workers report that they can obtain only 20 to 30 weeks of farm work annually. The dilemma of matching farmworkers and farm jobs to minimize unemployment for workers and crop losses for farmers has been debated for decades.

Unlike the "hired hands" of the Midwest, California farmworkers have usually been "aliens" in the sense that many are not U.S. citizens and "aliens" in the sense that they do not share social or cultural bonds with farm employers. Farmers who often could not communicate with succeeding waves of immigrant workers came to see farmworkers as homogeneous and distant entities: if one group of workers complained, the farmer simply asked for another labor contractor or crew. Thus, managing farm labor has been viewed by many farmers as similar to managing irrigation water: farmers' efforts center on assuring an overall surplus, so that when labor or water is needed, workers (or water) is readily available.

Farmers who could not communicate with immigrant workers turned the task of recruiting and supervising workers over to intermediaries. These intermediaries could have become worker advocates, but instead, most of them became labor brokers who allocated scarce jobs in a surplus labor market. Farm labor intermediaries are associated with some of the worst cases of worker abuse.

The most important intermediaries who match workers and jobs are bilingual farm labor contractors (FLCs) and field supervisors. FLCs are independent entrepreneurs who assemble one or several crews of 20 to 35 workers and arrange a succession of farm jobs for them. FLCs are paid or charge a fee for their services; typically 10 to 20 percent of what the farmer pays for FLC workers is kept by the labor contractor. In the mid-1980s, a citrus FLC with 100 pickers employed for six months could expect a profit of $30,000 to $40,000 for himself (almost all FLCs are men). Some FLCs operate only in one area; others take workers to other areas in order to maximize the period of employment for "their" crews. Some FLCs provide transportation, housing, food, loans, and work equipment; others provide only employment.

Countless stories of FLCs abusing workers through overcharges and underpayments have prompted repeated efforts to reform or extirpate FLCs. One summary of the FLC role in agriculture notes that many labor contractors are legitimate business people who provide a real job-worker matching service and that a labor contractor is often "a convenient whipping boy for the evils of the agricultural labor market."[17] However, the disadvantages of FLCs are thought to dominate: they exact a high cost from workers; workers are too vulnerable to exploitation; and too many contractors abuse workers in the "ill-defined and ambiguous nature of the employment relationship" between workers, labor contractors, and farmers.[18]

Field supervisors, the other major intermediaries, are often year-round employees who are responsible for recruiting and supervising workers during the busy harvest period. In addition to the FLC-supervisor system, some large companies hire workers directly or through labor co-ops, the public Employment Service matches farmworkers and jobs, and union hiring halls also refer workers to jobs.

The farm labor market in California became segmented or "balkanized" in the 1970s. Large unionized or union-influenced employers began to closely monitor the recruitment and supervision of workers to assure that supervisors applied the same hiring and firing criteria to forestall unionization or because a union contract required such procedures. However, small growers who did not fear union activity continued to permit supervisors or FLCs to handle workers without interference. This lack of employer oversight permitted the deterioration of working conditions, wages, and the dominance of illegals in some crews. Growers remain liable for the actions of their supervisors and labor contractors under federal and state labor laws, so that even if the farm owner instructs a supervisor or labor contractor not to threaten workers in order to persuade them to vote against the union, and despite the owner's instructions the supervisor threatens workers, the farm owner is liable.

Most farmworker recruitment is of the informal word-of-mouth variety. Job and wage information is transmitted from FLCs or supervisors to currently employed workers, who inform their friends and relatives in the United States and in Mexico of job vacancies.[19] These worker information networks are very efficient, capable of bringing additional workers illegally from Mexico within two to four days. Since farm wages vary within California agriculture, this job information network is a valuable asset for workers in Mexico who want work in the United States, as the network refers them, for example, to high-wage lettuce jobs, mid-level citrus jobs, and lower-wage hoeing jobs.

California farm wages vary by commodity, region, and size of employer. In July 1986, a USDA employer survey reported that California fieldworkers averaged $5.22 hourly, 19 percent more than the $4.39 U.S. average. California piece-rate workers averaged $6.57 per hour, 44 percent more than the U.S. average of $4.55.[20]

A more detailed 1982 employer survey reported that the average wage of year-round and seasonal workers in California was $5.16 hourly.[21] However, wages for seasonal workers in this survey varied from $3 hourly to $20 per hour. Wages were generally highest on fruit and vegetable farms, which often offer piece rate harvesting wages, and lowest on livestock farms. Regionally, wages were highest in the coastal valleys near the state's major urban areas and lowest in the sparsely populated Sacramento Valley. Although past surveys have not collected detailed wage data, anecdotal evidence suggests that the largest employers are most likely to offer higher wages because of union demands or to avoid unionization.

Hourly wages do not indicate annual earnings. A worker employed for fifty 40 hour weeks in 1982 at the $5.16 average hourly wage would have earned $10,320, about 30 percent less than the $13,350 that a nonfarm worker would have earned at the $267 weekly average nonfarm wage. However, few California farmworkers work year-round; committed farmworkers average about 26 weeks of work annually. *Farm Labor* reports that hired workers averaged 43 hours per week, so a $5.16 hourly wage for 26 43-hour weeks yields average annual earnings of $5,769 for adult men. Women and children tend to earn lower hourly wages and work fewer weeks, so they have lower annual earnings.

Most adult farmworkers earn $150 to $300 weekly for 15 to 30 weeks, yielding farm earnings of $2,250 to $9,000.[22] Some farmworkers supplement their earnings with nonfarm employment, and many draw up to half their farm earnings in unemployment insurance (UI) payments. In 1984, the average 229,000 crop and livestock workers covered by UI had gross earnings of $2.7 billion, or a worker employed year-round in a crop or livestock job could expect to earn $11,800.[23]

How will farm labor markets continue to evolve in California? Large vegetable and citrus farms in California's coastal valleys appear capable of recruiting and retaining legal Mexican and Mexican-American workers who earn relatively high wages for 20 to 30 weeks of farm work each year. However, most California farm employers are in the Central Valley, where most farms continue to offer short and low-wage jobs to an army of seasonal farmworkers. In the Central Valley, most employers prefer to maximize the pool of farmworkers available rather than adjust their farming operations or develop labor management systems to use fewer workers for longer employment periods. Fruit and vegetable agriculture is likely to shift from the high-wage and union-influenced coastal valleys to the Central Valley over the next two decades to take advantage of lower labor and land costs.

E. SUMMARY

California's labor-intensive agriculture has depended for more than 100 years on an army of seasonal workers to pick perishable commodities during critical harvest periods. California's farm labor market continues to offer seasonal jobs that American workers with other options tend to shun.

California farm labor markets have become balkanized because mechanization, immigration, and collective bargaining do not have the same effects on all farms or commodities. Employers and workers in some commodities and areas, especially large coastal valley vegetable producers, have the type of labor market that assures experienced workers high wages and unemployment insurance during the off-season. However, the farm labor markets of the Central Valley that employ almost half of California's farmworkers offer much lower wages and fewer hours of work. The Central Valley promises to become more important as agriculture shifts away from the high-cost coastal valleys.

Many Central Valley farms operate as they have for decades. Labor contractors and supervisors are the intermediaries who recruit and supervise workers, so working conditions and sometimes wages depend on the integrity of these key intermediaries. Wages and working conditions in the Central Valley vary considerably, and are generally inferior to those in the coastal valleys.

NOTES

1. Economic Indicators of the Farm Sector (Washington: USDA-ERS, 1987).

2. Average annual hired worker employment (1.3 million) as a percent of average total employment (3.8 million) in 1980; for California, July 1983 hired worker employment (257,000) as a percent of total employment (327,000) Farm Labor (Washington: USDA, NASS, quarterly).

3. This section is based on the chapter "California" in P. Martin (ed.) Migrant Labor in Agriculture: An International Comparison (Berkeley: Giannini Foundation, 1984).

4. Concentration ratios change with mergers and business failures; this data is drawn from a variety of industry publications for the mid-1980s. An excellent survey of the lettuce industry and work force is William Friedland, Amy Barton, and Robert Thomas. Manufacturing Green Gold: Capital, Labor, and Technology in the Lettuce Industry. (New York: Cambridge University Press, 1981).

5. Agricultural Employment Estimates, 881-M (Sacramento: Employment Development Department, ED&R, 1987). Unemployment insurance data indicate that in 1985 crop and livestock employment ranged from 189,000 in February to 312,000 in September for a peak-trough ratio of 1.6 (all workers, not just seasonals).

6. These monthly estimates are reported in monthly Farm Labor Reports published by EDD and summarized annually in Agricultural Employment Estimates op cit.

7. This total work force estimate is based on the number of workers or social security numbers in the unemployment insurance system who had at least one job with a crop, or livestock, or agricultural services employer in 1985.

8. Economic Indicators of the Farm Sector op cit.

9. Leslie Smith and Robert Coaltrane. Hired Farmworkers: Background and Trends for the Eighties (Washington: USDA, ERS, Report 32, 1981).

10. The total California farm work force was 600,000 to 700,000 in the 1960s. In the mid-1980s, about 1 million social security numbers (SSNs) had farm wages in California. It is not clear why this increase occurred: the possibilities include an upsurge in illegal aliens with multiple SSNs; rapid expansion of agricultural services; and the undercounting of (especially service) workers in the 1960s.

11. Migratory Labor in American Agriculture (Washington: Report of the Presidential Commission on Migratory Labor, 1954).

12. Average employment as estimated in EDD Agricultural Employment Estimates op cit.

13. For a review of trade in labor-intensive fruit and vegetable commodities, see the Proceedings of Marketing California's Specialty Crops, Conference Report (Davis: Agricultural Issues Center, 1987).

14. This section is based on R. Mines and P. Martin. A Profile of California Farmworkers (Berkeley: Giannini Foundation, 1986).

15. Specifically, 55 percent reported that they had "greencards" which permitted them to live and work in any U.S. job and, in five years, become naturalized U.S. citizens. However, the validity of these greencards was not established, and it is presumed that many greencards in use in the mid-1980s are forgeries.

16. Lloyd Fisher. The Harvest Labor Market in California (Cambridge: Harvard University Press, 1952).

17. Arthur Ross and Samuel Liss quoted in Stephen Sosnick. Hired Hands: Seasonal Farm Workers in the U.S. (Santa Barbara: McNally and Loftin, 1978), p. 99.

18. Ibid., p. 100.

19. R. Mines, The Evolution of Mexican Migration to the U.S.: A Case Study (Bereley: Giannini Foundation Report 82-1, 1982).

20. Farm Labor (Washington: USDA, NASS, 1986).

21. G. Johnson and P. Martin, "Employment and Wages Reported by California Farmers in 1982," Monthly Labor Review, September 1983, p. 27-36.

22. Mines and Martin, 1986 op cit.

23. Unemployment Insurance Weeks Compensated by Industry (Sacramento: EDD, ED&R Report 352, 1985).

Chapter 2

The Evolution of California Agriculture

The first U.S. Census of Population in 1790 found that 90 percent of all Americans lived in rural areas and most were farmers or farmworkers. Farming was "a way of life" as well as a business operation that supported farmers and their families. The dominance of family farms in the Northeast produced an agrarian philosophy that exalted agriculture. The French physiocrats had argued that land and agriculture were the true pillars of a society's wealth, and Thomas Jefferson promoted this agricultural fundamentalism by arguing that a system of family farms was healthy for the economy and essential for American democracy.

This agrarian philosophy helped to shape the values that made America unique. The single most important factor responsible for this unique American outlook was the availability of land. European immigrants were conditioned to the idea that land ownership conveyed wealth and status on the owner, so the availability of 160 acres to almost everyone reinforced "exceptionalism," the idea that America was unique and that Americans were a uniquely favored people.

The availability of land and the philosophy of agrarianism made diversified family farms the goal of American agriculture. The model farm contained a hard working farmer, his family, and perhaps a hired hand who was working for wages in order to save enough money to buy his own farm. There was an important exception to this family farm ideal: the large plantations based on slave labor in the Southeast. But it was the family farm philosophy that dominated attitudes toward agriculture and farm labor.

The few "hired hands" on northeastern and midwestern farms in the early 1800s were thought to be temporary additions to the farmer's family, that is, the hired farmworker lived and ate with the farm family. This idealized notion that a hired hand was a temporary occupation on the path to farm ownership became less and less credible as land prices escalated much faster than farmworkers' wages. Still, the myth persisted that farmworkers were future farmers whose aspirations were the same as their farm employers. Today it is clear that farmworker and farmer are two different occupations. Most farmers get into agriculture via inheritance or with money earned in a nonfarm job, not from money saved by being a hired farmworker.

A. THE SEASONAL FARM LABOR PROBLEM

The major farm labor problem in California has been and remains seasonality—most workers find farm work for only two to six months of every year. Farmers worry about the cost and availability of labor to perform critical seasonal operations, while workers worry about how long they will be able to do farm work, what they will earn, and what they will live on during the off-season. Historically, most seasonal farmworkers have had to fend for themselves when seasonal farm work is completed.[1]

Experts who have grappled with the seasonality issue usually end up advocating reforms. Varden Fuller urged the enactment of labor laws that would set minimum wages and working conditions in agriculture and establish a framework that allowed farmworkers to organize and bargain with employers for further improvements.[2] Other reformers urged farmers to restructure their farm operations to hire fewer workers for longer periods.[3] Since the 1960s federal and state governments have developed programs to assist migrant and seasonal farmworkers (MSFWs) and their children, first by helping MSFWs to leave agriculture and more recently by offering training that makes some field workers employable in agriculture year-round.

The seasonal farmworker problem persists because California agriculture "has no regular labor force of its own and does not compete in the general labor market. Instead, it relies on the residuals of other labor markets."[4] Usually, these "residual workers" have been immigrants or strangers in California without other job options. Since farmwork is a last resort, most workers (or their children) hope to escape the cycle of seasonal work for lower-than-average wages.

Farmers have traditionally argued that immigrant workers were necessary to do seasonal farmwork because "Americans" refused such jobs. An 1883 editorial in the Pacific Rural Press noted that "the (fruit) crop of the present year, although deemed a short one, taxed the labor capacity of the state...the labor is not now in the county to handle an increase in production." The editorial went on to note that California farmers cannot hope to employ Americans or European immigrants because farmers "cannot employ them profitably...more than 3 or 4 months in the year—a condition of things entirely unsuited to the demands of the European laborer."[5]

The theme that only particular groups of immigrant workers were suited to work in California agriculture recurs in the speeches of farm leaders. In the late 1920s, a farm spokesman praised Mexicans as the best farmworkers because a

Mexican "is the result of years of servitude (he) has always looked upon his employer as his padron." Furthermore, "crouching and bending (farm) operations must be performed in climatic conditions in which only the Orientals and the Mexicans are adapted."[6] During the 1980s immigration reform debate, the same arguments were heard: California agriculture "has had to rely largely on foreign workers to harvest the crops [...because] it has not been the type of work, unfortunately, that has attracted large numbers of domestic workers."[7]

For over a century farmers have argued that labor-intensive crops cannot be grown without an ample supply of immigrant workers, and that they have a right to special assistance from the government to obtain farmworkers.[8] Farmworker advocates have usually disputed the need for immigrant workers, arguing that farmers "needed" only to raise wages in order to attract American workers. This debate has persisted for a century, with farmers usually winning.

Farmers and farmworker advocates also debate wages and working conditions. California farmers hire over 1 million workers each year. Farms are often located at considerable distances from farmworkers' homes, giving rise to some migrancy and some migrant housing. Farmers are reluctant to build and maintain permanent housing for employees who are on the job only several weeks or months each year. Farmers are also reluctant to pay high wages to these seasonal farmworkers because they believe they cannot pass higher wage costs on to processors and consumers. Thus, farm wages remain lower than the U.S. average and farmworker housing is often substandard because there is usually a surplus of immigrant workers who are not in a position to bargain for improvements.

Farmworkers may promise individual farmers that they will return every year when needed, and many do. However, seasonal farm work is a last resort job for many farmworkers, and many hope that next year they will not have to do farm work. Also, individual farmworkers have little incentive to stay working for one farmer or in a particular farming area when they learn that better yields or better piece-rates are available elsewhere.

The seasonal farmworker problem can be summarized as one in which most farmers do not care who picks their crops and do not worry about what will happen to this year's work crew when the farm work is completed. Instead, farmers seek to maximize the pool of workers available to fill seasonal jobs. Meanwhile, most farmworkers do farmwork only as a last resort and hope to escape to higher wage or year-round nonfarm jobs.

B. THE EVOLUTION OF CALIFORNIA AGRICULTURE

California farm labor history is the story of how seasonal farm work emerged as a "last resort" for immigrant workers with few alternatives. The California farm labor market matches seasonal farmworkers and jobs every year, and historically, the wages and working conditions that immigrant farmworkers without options were willing to accept have largely determined wages for all farmworkers. As farm wages stagnated or rose slower than nonfarm wages, farmworkers or their children were attracted to nonfarm jobs that offered higher wages, better working conditions, and year-round work, and they were replaced by immigrants. This unstructured farm labor market dominated by immigrants created a large group of poor farmworkers and generated a lengthy literature on the plight of the farmworker.

The major labor question in the evolution of California agriculture is how California agriculture developed large farms growing labor-intensive fruits and vegetables while offering such low wages and part-year work in fast-growing California. The answer is that waves of immigrant workers willing to accept seasonal farm jobs were available, and each time a labor shortage threatened the structure of agriculture, another wave of immigrant workers arrived. The history of California agriculture can be divided into several periods. From 1769 to 1833, the 21 Spanish missions established along El Camino Real (Highway 101) employed Indians to work on surrounding farmland. This mission agriculture was not productive or efficient because, according to Fuller, a primary purpose of mission agriculture was to keep the American Indians at work in order to minimize disciplinary problems.[9] After the 1833 Secularization Act, private individuals could own 4,500 to 50,000 acres of California land in "ranchos" devoted to cattle grazing, and during this period, Indians were farmworkers. These Indian workers were "so cheap that growing crops were protected from cattle by Indian guards rather than fences."[10]

In 1850, California entered the United States as a nonslave state because the 1848 Gold Rush had brought enough anti-slave prospectors to California to tip the balance against slavery. When California entered the union, it agreed to recognize the Spanish land grants that had been made to individuals before the 1848 Treaty of Hildalgo, but confusion and corruption allowed a few wealthy settlers to acquire large tracts of the best land.

In the 1850s and 1860s, California agriculture consisted of cattle and sheep grazing on large ranches in the central and coastal valleys. However, the completion of the transcontinental railroad in 1869 and the demand for wheat in the eastern United States and in Europe encouraged large California farms to convert pasture into dryland wheat farms, establishing a tradition of crop specialization and monoculture in California in the 1870s.

Dryland agriculture required large acreages, but the acquisitive instincts of land buyers, plus the federal government's practice of giving land to the Southern Pacific railroad for laying track,[11] concentrated land ownership in the hands of a few. In the 1860s, a few thousand people held most of California's arable land. This concentration of land ownership persisted so that 70 years later 16 Californians each owned 54,000 acres.

1. Labor Choices in the 1870s

The 1870s were the critical decade that shaped the structure of California agriculture. Refrigerated transportation and irrigation were changing the economics of farming in California, making it profitable to grow labor-intensive fruit for distant markets. California farmers could continue to operate large acreages and minimize their need to hire labor by mechanizing the wheat harvest, they could subdivide their land into family-sized units, or they could develop an army of migrant farmworkers who would do seasonal farm work on large fruit and vegetable farms. There were only a handful of important large farmers making these decisions in the 1870s.

The preferred option of most farm organizations was to subdivide the state's large ranches and sell family-sized parcels to the midwestern farmers who were expected to arrive on the transcontinental railroad to develop family-sized fruit farms. However, the 12,000 Chinese who had built the railroad through the Sierra Nevada mountains were excluded from urban jobs and became employed as migrant workers on large and irrigated wheat and fruit farms. The Chinese proved to be ideal migratory workers from the farmers point of view: they organized themselves into crews, selected bilingual crew leaders, and handled housing and meals, so employers took little responsibility for them while they worked and no responsibility for them after the seasonal harvest work was completed.

The availability of Chinese labor just when California agriculture found it profitable to switch to labor-intensive fruit farms established patterns that have persisted for 100 years: crop specialization, large farms, and seasonal workers who migrated from farm to farm. Scale, specialization, and migrancy persisted because early California farmers were entrepreneurs who had the capital and the management skills needed to operate large and complex enterprises.[12] California farmers were not farmers in the traditional sense of living on their farms; many of them lived in San Francisco.

The availability of Chinese labor preserved large and specialized farms, institutionalized migrancy, and developed the farm labor contracting system that persists today. Few farmers spoke Chinese, so Chinese farmworkers organized into crews of 20 to 40 workers and negotiated with growers through a bilingual spokesperson. Initially, these intermediaries were simply other workers who understood both languages, but later they evolved into specialized farm labor contractors who recruited workers, supervised the harvest, and arranged housing and meals. While the Chinese and Japanese contractors tended to be the workers' advocates, Mexican labor contractors emerged in the 1920s with an independent business interest in maximizing the difference between grower costs and workers' wages. Contractors soon became as much "the enemy" as growers, and the object of many farm labor reforms.

By the 1880s, large irrigated fruit farms depended on migratory crews of farmworkers. The historical question is whether this transition from an extensive to an intensive agriculture without breaking up large farms was planned and manipulated by large farmers or whether the dependence on seasonal workers was an accident of large Spanish land grants and the availability of Chinese farmworkers. Regardless of the cause, once established in the 1870s, specialized fruit farms that depended on migratory farmworkers tended not to be converted into family farms of the midwestern type.[13]

Large and specialized California farms that depended on crews of migrant workers survived and prospered during the 1870s and 1880s. Many California landowners still expected to eventually subdivide their land and sell it to family farmers, but by the mid-1880s the large-farm and migrant worker system was well established.

In 1882, however, labor unions persuaded Congress to halt Chinese immigration. White workers opposed to "cheap" labor sought to round up and deport Chinese workers who were a majority of California's fruit and vegetable work force. In defense, California farmers formed associations to protect the rural Chinese from deportation campaigns.

In the mid-1880s, California farmers discussed and debated whether to reorganize agriculture and reduce their dependence on hired farmworkers or to seek a new supply of migrant farmworkers. Some farmers noted that diversified family farms would employ fewer hired workers year-round and bring stability to rural communities. The Pacific Rural Press in an editorial on February 11, 1888, noted that the end of Chinese immigration meant that California agriculture would have to change: "Farmers must now look to a normal population, for the age of the single male laborer will gradually slide away...diversified (farming) industry that will give some employment all the year-round is the solution. The married man cannot be a mover. He must have his cottage and garden, and a reasonable employment within a few miles. This is the normal condition of American life."[14]

Diversified farms and year-round employment remained unnecessary because an economic depression in the mid-1880s forced many Whites into the hired farm work force. Another recession in 1893 brought new waves of unemployed Whites into agriculture, and some destroyed Chinese housing in an effort to open up jobs for themselves. The labor supply question faded as California farmers concentrated on improvements in irrigation and fruit production and drying techniques.

By the end of the 1890s, farmers were once again complaining of labor shortages. This time, the remedy was another wave of immigrant workers: between 1900 and 1909, farmers imported 56,000 Japanese workers. But Japanese workers exhibited a propensity to organize and strike during the harvest season to demand higher wages. These "quickie" harvest strikes were extremely effective because other Japanese farmworkers refused to cross the picket lines set up by their striking countrymen and the Japanese collectively refused to work on certain farms. The Pacific Rural Press in 1909 wrote that Chinese farmworkers would be vastly preferred to Japanese and that only "enough Japanese to keep some lines of our agriculture going" should be imported. But, the article continued, "we do not wish too many, nor do we wish them to buy up and lease up all the goods things of the State and paint the future for Americans on this coast brown. We need those who will work for fair wages and fly away with them."[15]

The Japanese were the only immigrant farmworkers who managed to move from hired worker to owner status in significant numbers. Despite Alien Land Laws in 1913 and 1919 that prohibited non-U.S. citizens from buying California farmland or leasing land longer than three years, the Japanese were able to buy or lease and in the names of their American-born children and by 1920 dominate the production of labor-intensive berries, onions, asparagus, and green vegetables on 75,000 owned and 383,000 leased acres.

To minimize their dependence on the Japanese in the early 1900s, California farmers tried to recruit Whites from the East and Midwest by offering better housing and working conditions. These experiments largely failed because farmworkers recruited from others parts of the United States soon left the fields for California's cities, often before they repaid the farmers' transportation expenses. These midwestern recruits refused to keep themselves available for seasonal farm jobs when they could earn higher wages in year-round urban jobs, and their experience helped to firmly establish the status of California agriculture as a last resort occupation for American workers.

Other immigrant farmworkers arrived in California at the turn of the century, but in smaller numbers. Farmworkers from Europe did not suffer the discrimination directed against the Chinese and Japanese; and the Portuguese and Dutch soon owned dairies, the Italians and Yugoslavs owned vineyards, and the Armenians specialized in marketing fruits and vegetables. The goal of every immigrant group was to get out of hired farmworker status as soon as possible.

The Japanese and other immigrant farmworkers continued to work on the state's large and specialized farms until 1916, when World War I created a large number of year-round war production jobs. In 1917 and 1918, farm labor shortages resulting from the exodus to nonfarm jobs were remedied by the employment of Mexicans in California agriculture. Many of these Mexican farmworkers were not recruited actively by farmers but came north to escape the disruptions of the Mexican civil war.

2. Immigrant Workers After 1920

The Immigration Acts of 1921 and 1924 restricted immigration from Europe and Asia, but Mexico and other western hemisphere nations were not covered by quotas and Filipinos were allowed to come to the United States without visas until 1934.

As a result, during the 1920s the traditional residual farm labor supply was augmented by a rapid increase in the number of Mexicans and Filipinos. The Mexican population in California increased threefold in the 1920 to 368,000, while the Filipino population rose from 2,700 to 30,500. Mexicans soon comprised an estimated 50 to 75 percent of California's average 200,000 farmworker employment. The availability of Mexicans also permitted cotton production to expand rapidly in the Southeast, and the automobile facilitated migrancy. Mexican farmworkers were the first group of farmworkers to arrive in California with their families, thus making clearer the poverty of farmworker families. Mexicans also pioneered the concept of migratory farmworker families traveling from harvest to harvest in autos, further exposing farmworker conditions to the nonfarm public.

Mexicans soon became preferred farmworkers. A statement by a Chamber of Commerce spokesperson to Congress in 1926 illustrated common grower sentiments of the time:

> We, gentlemen, are just as anxious as you are not to build the civilization of California or any other western district upon a Mexican foundation. We take him because there is nothing else available. We have gone east, west, north, and south and he is the only man-power available to us.[16]

This persisting grower fear of harvest labor shortages and the assertion that only immigrants would do farmwork was tested during the 1930s Depression, which brought White farmworkers into the fields.

During the 1930s, the Depression brought Midwestern farmers to California. About 1.3 million people migrated from other states to California and at least 150,000 of these new Californians became migrant farmworkers. Throughout the 1930s, many California farmers and workers complained that too many workers were trying to find farm work, prompting the "repatriation" of 400,000 Mexicans. Some California farmers exacerbated the labor glut by advertising for two or three times as many workers as were needed, promising a (high) wage that was soon reduced when too many workers responded. The late 1930s influx of White farmworkers prompted a series of investigative reports which documented housing and working conditions, and these reports inspired John Steinbeck's book *The Grapes of Wrath*.[17]

In the 1940s, White farmworkers began leaving agriculture to take wartime industrial jobs or join the armed forces. The Japanese were placed in internment camps, and Filipino youth left agriculture for urban jobs or the armed forces. In the spring of 1942, California farmers called for the importation of between 40,000 and 100,000 Mexican farmworkers for the September harvest.[18] This alleged labor shortage was contested by American unions and by Mexican-American organizations; they argued that no imported farmworkers were needed. However,

1,500 Mexican farmworkers were brought to California in September 1942. Early in 1943, Congress passed PL45, the Farm Labor Supply Appropriations Act, which legitimized the "Bracero" program.

Congress enacted three more labor agreements with Mexico: PL229 in 1944, PL893 in 1948, and PL78 in 1951. Each of these "Bracero" laws was justified by farmer arguments that agriculture needed a supplemental immigrant work force. The peak number of braceros in California during the war years was 36,600 in August 1944, and the average number (26,000) represented about 20 percent of the seasonal work force. After the war, the Bracero Program expanded so that by 1956-57, a peak of 101,000 braceros were in California, and the average number (51,000) constituted 28 percent of the seasonal labor force.

The Bracero Program brought almost five million Mexican farmworkers to the United States before it was terminated in December 1964. Some braceros returned year after year, so only one to two million individuals actually participated. However, many braceros became legal immigrants—that is, they were born in Mexico but in the 1960s acquired the right to live and work legally in the United States.

Since the mid-1960s, California's hired farm work force has been composed of White adults (including women and teenagers); U.S. citizens and (green-card) immigrants of Mexican origin; illegal aliens or undocumented workers, especially from Mexico; and a variety of other immigrant groups, including Filipinos, Yemenis, Punjabis, and Vietnamese.

C. SUMMARY

California agriculture became addicted in the 1880s to a never-ending supply of seasonal and mostly immigrant workers because the large land units originally needed for dry land grazing and wheat farming were not broken up into diversified family farms. Large and specialized fruit and vegetable farms were preserved because they were managed by entrepreneurs with the capital and skills necessary to maintain large units and because a seasonal and migratory work force was readily available. California farmers soon came to believe that California agriculture could and should continue to be large scale, specialized, and dependent on hired farmworkers. The option of family farms offering better wages and working conditions to family workers faded before the assertion that the federal government should recruit or tolerate the presence of immigrant workers in California agriculture. By 1900, the framework of the farm labor debate was set by farmers who, in Fuller's words, "declared that California agriculture *by nature* was such as to demand a permanent supply of itinerant laborers. Since White people refuse to perform such 'menial tasks,' such a labor supply by its very nature had to be 'un-American'."[19]

Large and specialized California farms dependent on migrant farmworkers with few alternative job opportunities turned the farm labor market into a "residual" market, a last-chance employer for workers who were mostly immigrants. Given the available supply of workers, farmers did not have to develop expensive personnel systems to recruit, train and retain farmworkers. Instead, farmers encouraged the development of the crew leader or farm labor contractor (FLC) system in which an intermediary recruits and supervises farmworkers for a fee. Under the FLC system, workers are often paid piece-rate wages, minimizing supervisory responsibilities.

NOTES

1. A good summary analysis of the seasonal farm labor problem is in Varden Fuller and Bert Mason, "Farm Labor" in The New Rural America. Annals of the American Academy, Vol. 429, January   1977, pp. 63-80.

2. Varden Fuller. The Supply of Agricultural Labor as a Factor in the Evolution of Farm Organization in California. Unpublished Ph.D. dissertation, U.C. Berkeley, 1939. Reprinted in Violations of Free Speech and the Rights of Labor Education and Labor Committee, [The LaFollette Committee] (Washington: Senate Education and Labor Committee, 1942), pp. 19778-19894.

3. Donald Rosedale and John Mamer. Labor Management for Seasonal Farm Workers. A Case Study. (Berkeley: University of California Leaflet 2885, 1976).

4. Fuller, op cit., p. 19882.

5. Quoted in Ibid. p. 19813.

6. Quoted in Ibid p. 19868.

7. Congressman Leon Panetta (D-CA) in the Congressional Record, June 14, 1984, p. H-5839.

8. This "agricultural exceptionalism" is well-summarized in Robert Thomas, Citizenship. Gender. and Work (Berkeley: University of California Press, 1985) Chapter 2.

9. Fuller, op cit., pp. 19784-19795.

10. Fuller, Ibid., p. 19876.

11. The S.P. railroad received 12,800 acres for each mile of track laid (an acre is about the size of a football field).

12. Management and scale were needed, for example, to organize irrigation districts and to mechanize. Until 1887, all irrigation districts were organized privately—farmers could drill for water on their own property, but it was more efficient for few farmers to form an irrigation district and store water for irrigation. Mechanization was promoted by level farmland and predictable periods of rainfall.

13. Midwestern family farmers did not become California farmers for several reasons. Midwestern farmers sold their farms at relatively low prices, paid the cost

of moving to California, and then had to pay prices for farmland that reflected the profits obtainable by hiring migrant Chinese workers.

14. Quoted in Fuller, op cit., p. 19816.

15. Quoted in Fuller, op cit., p. 19827.

16. Quoted in Fuller, op cit., p. 19859.

17. These 1939 reports included the LaFollette Committee Hearings, op cit.

18. Richard Craig. The Bracero Program (Austin: University of Texas Press, 1971).

19. Quoted in Fuller, op cit., p. 19882.

Chapter 3

Farmworker Unions

California fields have witnessed struggles between organized farmers and unorganized workers for over 100 years. Successive waves of immigrant farmworkers called strikes to improve wages, but most of the documented history of union activity in agriculture refers to periods when recession pushed White workers into agriculture (1910-1917 and the 1930s) or the rise of the United Farm Workers in the 1960s. A brief review of early union activities illustrates the difficulties inherent in maintaining unity among seasonal farmworkers who want to escape from agriculture.

A. UNION ACTIVITY FROM WORLD WAR I TO 1965

The Industrial Workers of the World (IWW) or "Wobblies" was a revolutionary union committed to overthrowing the wage system in the early 1900s. IWW organizers were adept at getting jobs on farms, learning the workers' dissatisfactions, and organizing strikes to protest employer practices such as advertising for too many workers and then cutting wages when a surplus appeared; charging for drinking water in the fields; and requiring workers to live in and pay for substandard housing. The Wheatland hops riots of 1913 illustrate the employment practices that generated farmworker strikes and their aftermath.[1]

Wheatland is a small community 30 miles northeast of Sacramento that specialized in growing the hops used to make beer. Although Chinese and Japanese workers dominated the fruit, sugarbeet, and vegetable work forces, most workers in hops were migratory White men who travelled without their families. Following common practice, the Durst ranch advertised for 2,700 hops pickers, even though only 1,500 were needed. When 2,800 workers appeared, Durst offered a piece-rate wage of only $.90 per hundred weight, rather than the $1 going wage. In addition,

Durst required the workers to buy supplies from his relatives, including lemonade at 5 cents per glass. In typical IWW style, the workers formed a protest committee and presented a list of demands to Durst, including a demand for a wage increase. At a subsequent worker rally, law enforcement officials tried to arrest an IWW leader and caused a riot which killed a district attorney, a deputy sheriff and two workers.

Repression and reform quickly followed. IWW leaders were arrested, tried, and sentenced to long terms in Folsom prison. Governor Hiram Johnson appointed a commission on Immigration and Housing to recommend farm labor reforms, and the Commission urged housing improvements and an end to over-recruitment and wage cuts. The Commission blamed farm labor unrest on "bad living conditions and insecure and intermittent employment," but Varden Fuller noted that poor housing and seasonal employment had been integral features of the farm labor market for decades.[2] Fuller believed that labor unrest such as the Wheatland hops riot was due to urban unemployment which drove urban workers into the farm workforce who were not accustomed to the living and working conditions of the Chinese and Japanese.

During the 1920s, nonfarm businesses attempted to stave off unionization by developing formal grievance mechanisms to handle worker complaints and to restrict the activities of supervisors who dealt with workers every day. The U.S. government curtailed immigration, and many nonfarm employers began to realize that the threat of being fired was less effective to motivate workers than the offer of fringe benefits and written personnel policies. California farmers formed or joined commodity associations, such as the Western Growers Protective Association, and these associations began to establish commodity-wide wage scales and employment practices. Unlike nonfarm employers, who tried to develop employment policies which helped workers to identify with a particular employer or factory, the commodity associations encouraged farmworkers to identify with a commodity and not a particular grower in the 1920s and farmworkers began to lose touch with individual growers.

Mexican farmworkers in southern California organized several unions in the late 1920s to demand wage increases and to end the use of farm labor contractors (FLCs). One Mexican union called a strike in the Imperial Valley in 1928 and, although no contract was signed, growers agreed to end the practice of withholding 25 percent of each worker's wages until the harvest was completed and growers, rather than FLCs, became responsible for assuring that workers received their full wages.[3]

The 1930s produced the largest and most violent strikes in California agriculture. Commodity prices fell in the Depression, and rising urban unemployment drove nonfarm workers into agriculture. As competition between workers for farm jobs increased, farmers cut wages, by one estimate from 35 to 50 cents per hour in 1929-30 to 15 to 16 cents in 1933.[4] There were spontaneous farmworker strikes to protest these wage cuts.

The IWW had organized farmworkers who were protesting wage cuts and inadequate housing before World War I, and the Communist-dominated Cannery and Agricultural Workers Industrial Union assumed the same organizing role in the 1930s. The normal CAWIU practice was to assume the leadership of spontaneous protests, so that when workers "walked out" to protest e.g. a wage cut, CAWIU organizers would meet with the striking workers, help them prepare a list of demands, and then confront the employer. The CAWIU was involved in 10 such strikes around California between 1930 and 1932; most CAWIU members were Mexicans.[5]

The year 1933 was the high-water mark of farm labor strikes and CAWIU influence. During 1933 there were 37 strikes involving 48,000 farmworkers in 14 crops, culminating in an October cotton harvesters strike. Cotton farmers met every September to set a standard piece-rate wage for the October harvest, and in 1933, growers agreed to pay 60 cents per hundred pounds of cotton picked. This rate was up from 40 cents in 1932, but well below the $1 of the late 1920s.

The CAWIU had won considerable support among farmworkers for successfully winning wage increases in earlier 1933 strikes, so workers were inclined to listen as CAWIU organizers railed against the organized cotton growers. The strike was called, and growers immediately evicted strikers from grower-owned housing, pushing workers into CAWIU-run tent camps and promoting worker solidarity. Roving pickets soon confronted law enforcement personnel and grower-organized vigilantes, and there was violence.

Strikers ranks remained firm for several weeks, and the threat of the loss of the 1933 cotton crop led to a Governor's Committee to settle the strike. A compromise piece-rate of 75 cents was established, and banks persuaded farmers to accept it while the state persuaded workers to accept it by discontinuing emergency food relief.

The CAWIU campaign of 1933 was a success which scared farmers into action. The newly-formed Associated Farmers persuaded most rural counties to enact anti-picketing ordinances and a general strike in the summer of 1934 in San Francisco generated a "Red Scare" which led to the arrest of CAWIU leaders. CAWIU influence quickly melted after its leaders were imprisoned. The CAWIU demonstrated to growers that wage cuts would provoke worker discontent and damaging strikes, but the CAWIU's policy of not compromising or signing agreements had the boomerang effect of further organizing growers into an effective counterforce to confront unorganized workers.

The 1935 National Labor Relations Act (NLRA) granted organizing and collective bargaining rights to most nonfarm workers in the private sector, but farmworkers were excluded because agriculture was considered a unique industry.

Farmers argued that their annual incomes depended on their ability to harvest perishable crops within a few weeks, so that a union of seasonal farmworkers could wield enormous bargaining power at harvest time. Even though seasonal farmworkers who went on strike would lose several critical weeks of earnings, farmers won the argument and farmworkers remained excluded from the federal NLRA. Farmworkers were also excluded from other New Deal labor measures, including the Social Security Act, Unemployment Compensation, and the Fair Labor Standards Act.

Although the NLRA excluded farmworkers, the American Federation of Labor (AFL) began an organizing campaign among agriculturally-related workers in the mid-1930s. The Teamsters and Longshoremen's unions which transported farm commodities had the power to keep a struck farmer's commodities from going to market, and they organized field and packingshed workers. Since many farmers were vertically-integrated grower-shippers, counterpart vertical unions were deemed necessary. Both the Teamsters and a coalition of Maritime unions began to "march inland" in the mid-1930s, but inter-union rivalry and the skepticism of non-White farmworkers limited these vertical organizing campaigns.

During the late 1930s, there was a three-sided conflict for farmworkers between the Teamsters and other AFL unions, the rival United Cannery Workers of the Congress of Industrial Organizations (CIO), and the newly-revived Associated Farmers. Packingshed and cannery workers, dairy workers, and many produce truckers were organized into unions, but few fieldworkers were organized. Farm labor protest activity was quieted in the late 1930s by Associated Farmers measures to exploit the rivalry between AFL and CIO unions and because the arrival of Dust Bowl refugees, many of whom had been anti-union small farmers in the midwest, increased the farm labor surplus.

The 1930s closed with a spate of farm labor literature that described the plight of the farmworker and criticized farm employers. Two 1939 books had an enormous influence: John Steinbeck's *The Grapes of Wrath* and Carey McWilliams' *Factories in the Fields*. These books and the U.S. Senate's LaFollette Committee's 28 days of public hearings on farm labor conditions in December 1939 and January 1940 yielded a vast literature on California farm labor. The LaFollette report concluded that:[6]

> The economic and social plight of California's agricultural labor is miserable beyond belief. Average annual earnings for agricultural laborers are far below the minimum standard necessary even for the maintenance of an existence on proper levels of health and decency. Agricultural laborers are ill-fed, ill-clothed, poorly housed, and almost completely lacking in many other things commonly considered necessary

for civilized life. They have no job security, and except in rare instances no job preference or seniority.

The California agricultural laborer is underemployed and frequently meets the unfair competition of child and relief labor. He has no control over wage rates and no voice in fixing them. He must be housed for the most part in private labor camps dominated by the employer. He lacks adequate medical attention. His children are unable to secure satisfactory continuous education. He has no adequate protection from industrial accidents and no workmen's compensation. State minimum-wage and maximum-hour laws do not give him any protection.

Residence requirements often bar him from relief. Organized protests on his part have been met with the blacklist, the denial of free speech and assemblage through the application of illegal ordinances of various kinds and through acts of outright vigilantism. His right to organize and bargain collectively is unprotected.

The LaFollette Report was critical of large farm employers, especially their ability to create organizations which pressured law enforcement agencies to interfere with labor organizing. The LaFollette Report concluded that these "farmer organizations" really represented only the largest corporate farms:[7]

The business interests dominant in the agricultural department of the State Chamber of Commerce were the great corporate employers long dependent upon a docile oversupply of farm labor. They could conceive but one objective: the suppression by all available means of the unrest among agricultural employees. The most effective weapons for such a campaign of suppression were obviously the agencies of State and local law enforcement. The agricultural interests, therefore envisaged a campaign to bend these agencies to their purposes, a "red" hunt and a "red" scare to prejudice public opinion against protests lodged by the workers or their leaders, a chain of citizens' committees to influence local law-enforcement officers and support them in their activities, and a centralized direction to coordinate activities throughout the State and to track down the radical union organizer wherever he might go.

Labor organizing slowed as the United States prepared for World War II, largely because union activities which caused losses in production were considered "unpatriotic." As farmworkers found nonfarm jobs or entered the Armed Forces, farmers complained of labor shortages. Initially the farmers were rebuffed, but in August 1942 the U.S. and Mexican governments concluded an agreement to import

Mexican workers (braceros) to do farmwork in the United States. Unions complained that the growers' cries of labor shortages were "a mere repetition of the age-old obsession of all farmers for a surplus labor supply,"[8] but farmers won the right to import braceros by arguing that crop losses caused by labor shortages would hamper the war effort.

The Bracero Program began as a small wartime "emergency" guestworker program. After World War II ended, farmers alleged that they continued to face labor shortages, reviving the 1920s arguments that Americans won't do farmwork and that, without immigrant workers, crops would be lost and consumers would face higher food prices. As the Bracero Program expanded, its contradictions became evident, as a parody of Alice in Wonderland illustrates:

> "What are those people doing?" asked Alice, surveying a vast, fertile southwestern valley.
>
> "They're cultivating surplus cotton and lettuce" replied the Red Queen.
>
> "Who are they?" asked Alice...
>
> "They are Mexicans imported because of the labor/shortage," explained the Red Queen.
>
> "Labor shortage?" asked Alice, "I thought we had 5 million unemployed and a million or so migrant farmworkers who need work."
>
> "Obviously," retorted the Red Queen, "you don't understand the American agricultural system."[9]

The Bracero Program expanded in the 1950s, farm wages fell further behind nonfarm wages, and farmworkers who could find nonfarm jobs left agriculture. The AFL-chartered National Farm Labor Union (changed to the National Agricultural Workers Union [NAWU] in 1952) continued to organize California farmworkers, but with little success. Ernesto Galarza became the driving force behind the NAWU, and he believed that farmworkers could never mount effective strikes if legal Mexican braceros and illegal Mexican workers were available. Galarza campaigned tirelessly to end the Bracero Program, and protested bitterly when braceros were used (illegally) to break NAWU-called strikes.[10]

Farmers usually evicted striking workers from grower-owned housing and then "borrowed" braceros from neighbors. The NAWU would lodge protests, but by the time government agencies agreed that braceros should not have been used as strikebreakers, the harvest was usually completed. The AFL-CIO chartered yet another farmworker union in 1959, the Agricultural Workers Organizing Committee (AWOC), and the AWOC was expected to "do the job at last" and organize California farmworkers. The AFL-CIO committed organizing funds, put AWOC headquarters in Stockton, and expected the agricultural workers union to soon have two million members, which would have made it the largest union in the

United States. However, the AWOC was directed by traditional English-speaking organizers who attempted to organize the workers who were hired through daily "shape-ups" in the skid-row areas. The AWOC never became the nation's largest union, but it did help to raise the wages of Filipino grape pickers and the White "fruit tramps."

Unions made persistent efforts to organize farmworkers before the 1960s, but no union achieved lasting collective bargaining agreements with farm employers. Periods of past union activity coincided with sharp wage cuts: the Industrial Workers of the World (IWW) was active from 1910 to 1917 and the Communist-dominated Cannery and Agricultural Workers Industrial Union (CAWIU) from 1930 through 1933. The Agricultural Workers Organizing Committee in the 1950s became a Filipino-dominated organization that merged with the National Farm Worker Association into what eventually became the United Farmworkers Union (UFW) in 1967. The UFW, an organization headed by Cesar Chavez and dominated by Mexican-Americans, is the largest farmworker union in California and the United States.

Pre-UFW efforts to organize farmworkers failed for a number of reasons: the usual surplus of farmworkers was augmented by braceros and illegal aliens; the short harvest season allowed farmers to recruit strikebreakers and not have to deal with "labor troubles" until the next season; many farmworkers were unsure whether the FLC, the grower, or an employer association was their employer; and the tendency of unions to rely on nonfarm or inappropriate organizers left pre-1960s unions vulnerable to charges that "outside agitators" were stirring up trouble in rural communities. Farmers sometimes raised wages during the heat of a strike, but rarely signed contracts that required a continuing relationship with a fieldworker union. One exception was Bud Antle, a Salinas lettuce grower who in 1961 signed a contract covering fieldworkers with the Teamsters who already represented packingshed and transportation employees. Antle also obtained a $1 million loan from the Teamsters pension fund.

B. CESAR CHAVEZ AND THE UNITED FARM WORKERS (UFW) UNION

Cesar Chavez, a community organizer in San Jose and former farmworker, moved to Delano in 1962 to organize farmworkers. The first convention of the Farm Workers Association (FWA), which Chavez founded, was held in Fresno on September 30, 1962, and the FWA offered to help its members with government paperwork and access to a credit union in exchange for $3.50 monthly dues. In 1964, the FWA was renamed the National Farm Workers Association (NFWA), and in 1965 the NFWA conducted a successful rent strike against the public housing authority in Tulare County.

1. The UFW: 1965-1975[11]

In May 1965, Coachella table grape growers prompted a strike by Filipino farmworkers who belonged to the Agricultural Workers Organizing Committee (AWOC) by paying PL414 braceros $1.40 per hour while paying the Filipinos only $1.25. The AWOC called a strike, and even though grape growers refused to recognize the AWOC as bargaining agent for the Filipino workers, wages were raised to $1.40.

When the grape harvest moved to Delano in September 1985, grape growers offered $1.25 hourly plus a piece-rate of 15 cents per box. The AWOC called a strike, and some growers locked the striking Filipino workers out of the labor camps that were their winter homes. The NFWA also declared a strike and sent letters to growers demanding wages of $1.40 plus 25 cents. Farmers harvested the 1965 grape crop with FLC-supplied workers.

The NFWA called a boycott of Schenley Industries Scotch and Bourbon before the 1965 Christmas season because Schenley was a major California grape grower. In March 1966, Harrison Williams brought a Senate Subcommittee to Delano, and Robert Kennedy had a televised argument with the Kern County sheriff that generated national publicity for the boycott. In March and April 1966, after the NFWA marched to Sacramento to generate more publicity, Schenley agreed to negotiate a contract with the NFWA for $1.75 hourly plus 25 cents per box harvested.

The NFWA then switched to a boycott of DiGiorgio products, another large grape grower, and DiGiorgio responded by announcing that it had signed a contract with the Teamsters. This contract was later nullified and an election held to determine which union should represent DiGiorgio workers. In the August 30, 1966 election, the Teamsters won the right to represent packingshed workers, and the United Farm Workers Organizing Committee (UFWOC)—a merger of the AWOC and NFWA—won the right to represent DiGiorgio fieldworkers. On July 21, 1967, the renamed UFW and the Teamsters signed a jurisdictional agreement that left fieldworker organizing to the UFW and packingshed organizing to the Teamsters. The DiGiorgio contract was short-lived; in 1968, DiGiorgio sold its farmland and terminated the contract.

During 1967 and 1968, the UFW secured contracts for grape workers at major wineries such as Almaden and Gallo by threatening consumer boycotts. By 1969, the UFW had 12 contracts covering 5,000 wine grape growers.

After its initial success, the UFW sent letters to table grape growers in January 1968 requesting that they recognize the UFW as the bargaining agent for their fieldworkers. The table grape growers did not respond, and the UFW launched a national boycott campaign that urged consumers not to buy California grapes. Throughout 1968 and 1969, the UFW obtained endorsements of the grape boycott from unions, religious leaders, students, and urban politicians. Pressure on the grape growers increased as per capita grape consumption fell from 4.3 pounds in 1966 to 2.4 pounds in 1971.

In March 1970, Coachella table grape growers recognized the UFW and signed contracts that offered a $1.75 hourly wage plus a piece-rate of 25 cents per box, a union hiring hall, pesticide protections, and grievance procedures. In July 1970, 23 Delano table grape growers signed UFW contracts, so that UFW contracts covered about 20,000 jobs on 150 ranches, or almost one-quarter of the peak table grape workforce.

Between 1970 and 1973, UFW fortunes rose and fell. The UFW sent letters to Salinas lettuce growers in 1970 requesting recognition as the bargaining agent for their fieldworkers, and many of the lettuce growers responded by signing contracts with the Teamsters (in violation of the 1967 UFW-Teamsters agreement). The UFW managed to win contracts covering 15 percent of the lettuce workers, but the Teamsters soon represented the workers employed by growers who produced 70 percent of California lettuce.

By March 1973, the UFW had 180 contracts covering 40,000 jobs and claimed 67,000 members who were employed at least one day on 500 California farms. By the end of 1973, the UFW was down to 14 contracts covering 6,500 workers. Meanwhile, the Teamsters ended 1973 with 305 contracts covering 35,000 fieldworkers. In the spring of 1973, table grape growers in the Coachella Valley and Kern County decided to sign Teamster contracts instead of renegotiating the expiring UFW contracts. Gallo, which had a UFW contract from 1967 to 1973, joined the Delano table grape growers in switching to Teamster contracts in August 1973.

Throughout 1973, AFL-CIO leaders pressured the Teamsters to leave fieldworkers to the UFW. The Teamsters, who were not affiliated with the AFL-CIO, negotiated a tentative jurisdictional agreement with the UFW during the fall of 1973, but the agreement unraveled and the Teamsters stepped-up their farmworker organizing efforts in 1974. Thus, on the eve of the Agricultural Labor Relations Act (ALRA) in 1975, the Teamsters had largely replaced the UFW as the major union representing California farmworkers.

The UFW-Teamster rivalry was widely publicized in the early 1970s. Sosnick reviewed the rivalry between the Teamsters and the UFW and concluded that the Teamsters did not want another union to be in a position to call a strike that might put Teamster truck drivers and packingshed workers out of work.[12] Teamster dues of $8 monthly (two hours pay after 1976) increased the union's treasury, and the Teamsters feared that the UFW might eventually try to move from the fields into the Teamster-dominated packingsheds. Jimmy Hoffa, who was challenging Teamster's President Frank Fitzsimmons for control, was bitterly anti-UFW. Finally, the Teamsters resented being portrayed as the "bad guys" in farm labor disputes. Growers allegedly favored the Teamsters because it was a "business union" that did not press for a union-run hiring hall, veto power over mechanical harvesting equipment, and the regulation of pesticides.

2. The ALRA and ALRB in 1975

Edmund Brown, Jr. won the November 1974 California gubernatorial election and promised during his campaign to enact a farm labor relations law. Food and Agriculture Director Rose Bird wrote a compromise farm labor bill that won the eventual support of the UFW, growers, Teamsters, and other unions. Governor Brown called a special legislative session to consider the ALRA, and it was quickly enacted by the legislature. The ALRA was signed on June 5, 1975 and became effective on August 28, 1975, in time for the first elections to be held during September's peak harvest employment.

There was a rash of elections in the fall of 1975. Of the 382 elections held between August 28 and December 31, 1975, the UFW was certified in 191 elections (76 percent of all certified elections), "no union" was certified in 10 elections (4 percent of all certified elections), and the Teamsters and other unions, such as Fresh Fruit and Vegetable Workers and Christian Labor Association, were certified in 49 elections (20 percent of certified elections). The Teamsters withdrew from other elections before certification because of the jurisdictional pact with the UFW.

This substantial election activity depleted the budget of the Agricultural Labor Relations Board (ALRB), the agency created to enforce the ALRA. State senators unhappy with early ALRB decisions and regulations blocked a supplemental appropriation, forcing the ALRB to cease operations for five months. During this hiatus, the UFW qualified Proposition 14 for the November 1976 ballot, an initiative that would have amended the California Constitution to require the legislature to fund the ALRB. Proposition 14 lost—only 38 percent of the voters supported it.

In March 1977, the UFW and the Teamsters signed a five-year jurisdictional agreement that kept the Teamsters from organizing fieldworkers (except for the Bud Antle contract first signed in 1961), while the UFW agreed not to organize truck drivers and nonfarm packingshed workers. This pact was renewed annually from 1982 through 1985.

After the ALRA was enacted in 1975, the UFW became the dominant union representing fieldworkers. The UFW had about 200 organizers in the mid-1970s and 60,000 to 80,000 members. However, the UFW strategy soon shifted from organizing farmworkers to documenting unfair labor practices and urging the ALRB's General Counsel to vigorously enforce the ALRA. The UFW believed that decisions in the political arena would continue to have important effects on farmworkers, so the UFW became a major contributor of time, money, and printed matter to state and federal politicians who favored its causes. The ALRB, consequently, became a political football. Both the UFW and farmers had their legislative friends denounce particular ALRB actions and decisions rather than accept ALRB rulings as the best judgement of an impartial tribunal.

In January 1979 the UFW called a strike against major lettuce companies in the Imperial Valley to support its demand for wage increases, including a request that the general farm laborer wage rate be raised from $3.70 to $5.25 hourly, an increase of 42 percent. The 28 affected lettuce growers resisted, arguing that President Carter's 7 percent guideline for wage increases prevented them from considering any wage demand above 7 percent. The growers hired a public relations firm to encourage striking workers to return to work and threatened to make citizens' arrests of UFW picketers. Picketers around the fields where strikebreaking workers were harvesting lettuce and the private security guards hired by growers soon clashed, and a UFW picketer was killed. Poor weather and the strike combined to triple lettuce prices.[13] Growers not affected by the strike (because they had contracts with other unions) reaped windfall profits, and some of the struck growers who were able to harvest part of their lettuce with strikebreaking crews obtained extraordinary profits because of higher lettuce prices.

The lettuce strike had not been resolved when the harvest moved northward to Salinas in the summer of 1979. The UFW threatened to boycott Chiquita bananas to increase its pressure on Sun Harvest, the United Brands subsidiary that harvested lettuce. The UFW charged that law enforcement officials were siding with the growers by arresting picketers and ignoring growers who imported strikebreaking illegal aliens. Growers charged that roving caravans of pickets assaulted strikebreaking workers to drive them out of the fields.

The strike continued throughout the summer of 1979, until the Teamsters renegotiated the contract with Bud Antle for a base wage of $5.00 hourly. The UFW soon negotiated $5.00 per hour contracts with two Salinas tomato growers, and near the end of August, West Coast Farms became the first lettuce company to negotiate a $5.00 general field wage with the UFW. A Sun Harvest agreement followed on August 31, 1979, and in September most of the remaining Salinas vegetable companies agreed to three-year contracts.

The 1979 Sun Harvest agreement, covering 1,200 vegetable workers, became the standard UFW contract. The Sun Harvest agreement included: ALRA good standing, which permits the union to be the sole judge of the good standing of its members; recruitment through a union-run hiring hall, binding arbitration over mechanization disputes; full-time company-paid union representatives; and automatic cost-of-living wage increases.

In July 1980, the UFW achieved a three-year contract with the Vintner Employers Association that represented 1,500 grape harvesters employed by Almaden, Paul Masson, Minstral, and Las Colinas vineyards. The vintner's contract increased general laborer wages from $3.80 to $5.10 (34 percent) and stipulated that vineyards harvesting all of their wine grapes by hand could machine harvest no more than 30 percent of their acreage by 1983. A three-week garlic workers' strike in San Benito and Santa Clara counties in July 1980 ended with an increase in the minimum hourly wage from $3.00 to $4.00 and a piece-rate of $2.50 per basket.

The UFW claimed 105,000 members at its peak in 1981, even though a September 1981 UFW report discussed only 30,000 "regular dues-paying members." When questioned, UFW officials could not reconcile the discrepancy.

After the 1981 peak, UFW fortunes suffered. When the lettuce harvest moved south in 1981, UFW members returned to work on the ranches still without contracts. The UFW stepped up its boycott of Bruce Church's Red Coach lettuce to persuade Bruce Church to sign an agreement, and early in 1984 Lucky Stores announced that it would no longer handle Bruce Church lettuce. However, in late 1987, Bruce Church has still not signed an agreement. Most of the lettuce growers who produced lettuce only in the Imperial Valley did not renew their UFW contracts, and some of the vegetable companies that had signed UFW contracts went out of business, including Sun Harvest.

The UFW aggressively pursued charges and complaints that the Imperial vegetable growers had bargained in bad faith since 1979, and that workers were entitled to make-whole wages. A March 1984 ruling by a San Diego Appeals court overruled the ALRB finding that the 28 lettuce growers had engaged in bad faith bargaining and concluded that the growers were simply engaged in hard bargaining. The UFW suffered another blow in 1987 when an Imperial Valley judge ruled that the UFW was responsible for the violence at one ranch during the 1979-80 vegetable strikes and owed $1.7 million in damages. The UFW argued that this damage award would effectively put it out of business, although the UFW raised enough money to post a bond and appeal the decision during the summer of 1987.

Thus, between 1976 and 1984, farmworker activity shifted from the fields and picket lines to ALRB hearing rooms and courtrooms. Instead of televised picket line confrontation between strikers and strikebreakers, farm labor disputes became legal arguments over why the union or employer took a certain action or whether the parties were really bargaining to reach an agreement.

After George Deukmejian was elected governor in 1982 and began to replace the members of the ALRB, the UFW alleged that the ALRB had become pro-grower, and threatened to resume its strike, picketing, and consumer boycott activity. The UFW began a grape boycott during the summer of 1984, urging consumers not to buy California table grapes because they contained pesticide residues. The UFW also began a boycott of the ALRB and the UFW tried to have the California Legislature block the ALRB's funding.

3. Farmworker Unions in the Mid-1980s

In 1987, six farmworker unions claimed 22,000 to 26,000 farmworker members on about 334 farms, or less than 3 percent of the one million individuals who do farm work for wages sometime during the year. Some of these farmworkers are employed on a farm just one day and may be union members just for that day. Farmworker unions represent only a small fraction of the workers on crop and livestock farms.

The UFW has the most ALRB certifications and probably the most farm jobs covered by union contracts (Table 3.1), but the Christian Labor Association has almost four times as many contracts. The Christian Labor Association and Teamsters Local 63 have 235 contracts with southern California dairies, and these dairy contracts, each covering an average of four farm jobs, account for almost 70 percent of all union contracts in California agriculture.

The United Farm Workers claims 50 to 60 fieldworker contracts throughout the state, a sharp drop from the 115 contracts claimed in 1984. Many of the 1987 contracts are with nurseries, mushroom farms, and other specialty crop growers; the UFW has lost many of its "farm" contracts. The union is divided into four commodity divisions and reported in 1984 that it had 40 contracts in grapes and tree fruits, 36 in agricultural specialties such as nurseries and mushrooms, 22 in vegetables, and 17 in citrus. These reported contracts included farms with a decertification vote or expired contracts, if the union alleges that improper voting led to the decertification or unlawful bargaining prevented the negotiation of a contract.[14]

The UFW claims 20,000 to 40,000 members, but UFW membership figures must be interpreted with caution. The UFW, like other affiliated unions, pays dues to the state and national AFL-CIO organizations based on membership. The UFW paid dues in the mid-1980s to the national AFL-CIO for 12,000 members and to the California AFL-CIO for 5,000 members. It is not clear how these 5,000 and 12,000 member reports to the state and national AFL-CIO square with the UFW's claimed larger membership. In a June 1987 interview, Cesar Chavez claimed 30,000 to 35,000 members (down from a peak of 100,000 in the mid-1970s), an organization with assets and investments of almost $6 million and annual dues income of $1.9 million, and a union preparing for elections on farms after the ALRB leadership is replaced.[15]

Teamsters Local 890 has eight Salinas-area contracts, of which the major contract is with Bud Antle. The Salinas-based Independent Union of Agricultural Workers (IUAW) has five contracts covering about 1,200 jobs. In 1984 the

Table 3.1
California Farmworker Unions, 1987

Union	Members	Elections Certified	Commodities	Contracts	Jobs	Regions
UFW Keene CA 93531	6,000-10,000	370	Vegetables Horticulture, Citrus Grapes and Tree Fruits	50-60	5,000	Statewide[2]
Teamsters 890[1] 207 Sanborn Road Salinas CA 93901	12,000	10	Lettuce Mixed Vegetables Nursery; Eggs	8	4,000	Statewide
Independent Union of Agricultural Workers Box 5519 Salinas CA 93905	12,00-1,400	5	Mixed Vegetables	5	—	Salinas Central Valley Imperial Valley
Fresh Fruit and Vegetable Workers Local 788 471 Main Street El Centro CA 92243	1,800	16	Lettuce Coolers Vegetable and Melon Packing Sheds	16	—	Imperial Valley So. California
Christian Labor Assoc. Local 17 14997 Euclid Avenue Chino CA 91710	580-620	200-210	Dairy	210	600	Central Valley So. California
Teamsters Local 63 1616 W. 9th Street Los Angeles CA 90015	250	35	Dairy	35	40	Chino Central Valley
Total	22,000-26,000	646	—	334	9640	—

Source: The unions and industry groups

[1] WCT was certified by additional elections, but withdrew from them after the 1977 UFW-WCT Agreement.

[2] A telephone survey indicated two vegetable contracts and 1,500 members in Calexico; two grape contracts and 9,00 members in Delano; 17 contracts (12 in apples) and 1,5000 members in Watsonville; six contracts and 1,000 members in Salinas; 12 contracts and 1,300 members in Ventura; and three contracts and 130 members in San Diego. The Santa Maria based International Union of Agricultural Workers was dissolved in November 1985. In 1984, it had been certified in 32 elections and reported 2,000 members and 28 contracts.

International Union of Agricultural Workers, based in Santa Maria, had 28 contracts covering 1,400 jobs, but this union was dissolved in November 1985. The Fresh Fruit and Vegetable Workers Union (FFUW) Local 788, in the Imperial Valley, has 16 contracts, which cover both fieldworkers and packing shed workers.

Most California farmworker unions have had unstable relationships with farm employers. The longest continuing union-employer relationship is the Antle-Teamsters Local 890 agreement first signed in 1961. The Christian Labor Association was a mutual assistance organization which helped the two to 10 workers at various southern California dairies obtain at least minimum salaries and fringe benefits; the CLA became a union and signed contracts after the ALRA was enacted. The Independent Union of Agricultural Workers and the Fresh Fruit and Vegetable Workers have been under stress as many large vegetable growers turned to labor contractors and these unions experienced leadership changes.

The UFW remains the dominant farmworker union in the eyes of most Californians, but has changed its strategy several times as it has evolved. A wave of decertification votes and farm sales has sharply reduced the number of UFW contracts, since new owners often change crops and work forces. The UFW signed several agreements in 1986 with southern California vegetable growers at wages substantially below its original demands, reportedly to stem the loss of members. The UFW appears to be strongest in 1987 in horticultural specialty operations such as nurseries; the only commodity which is mostly harvested by UFW is mushrooms. The UFW has not been able to replace many of its lost vegetable contracts, and may have only 50 to 60 contracts and 6,000 to 7,000 members in 1987.

4. Summary

Farm labor conflict is a legacy of large California farms that depend on armies of seasonal workers. The unstructured farm labor market in which farmers tried to maximize the pool of workers while farmworkers scrambled for jobs generated few lasting employer-employee relationships and a great deal of worker resentment over wage cuts and inadequate housing. However, the "outsider" status of most union organizers and the usual surpluses of immigrant farmworkers kept worker protests and strikes from generating lasting collective bargaining agreements.

Cesar Chavez and the UFW marked a new era in farmworker organizing. Chavez was an effective organizer in the Civil Rights era when Americans were mobilized to fight for social justice. However, the UFW has had a roller-coaster existence since 1965; it peaked in 1970, almost disappeared in 1973, made a comeback after the ALRA was enacted in 1975, and has been shrinking since 1980.

Major farm labor issues since 1975 have centered around farmworker unions and the ALRB. Recently, union activity has become a less important farm labor issue than immigration reform, while tax, trade, pesticide, and commodity program issues have supplanted labor issues in the agricultural policy arena. Farmworker organizing and negotiating activity has virtually stopped—in 1985-86, there were very few elections and contracts negotiated or renegotiated.

Farmworker unions generally and the UFW in particular are seeking ways to attract new members and to negotiate contracts. The UFW is emphasizing the dangers of pesticides to workers and consumers, and the importance of the union to legalize farmworkers under the Immigration Reform and Control Act (IRCA) of 1986. The pesticide dangers are the basis for the UFW's appeal to consumers, who are being urged to boycott California table grapes. The IRCA's provisions for legalizing illegal alien farmworkers who worked at least 90 days in perishable agriculture between May 1, 1985 and May 1, 1986 offer the UFW an opportunity to win the loyalty of an important group of farmworkers if the UFW is the organization which acts as an assistance agent for such workers.

NOTES

1. Stuart Jamesion. <u>Labor Unionism in American Agriculture</u>. (Washington: U.S. Bureau of Labor Statistics, Bulletin 836, 1945), pp. 60-63.

2. <u>Ibid</u>, p. 63.

3. <u>Ibid</u>, p. 77.

4. <u>Ibid.</u> p. 80.

5. <u>Ibid</u>, p. 122.

6. "Violation of Free Speech and Rights of Labor," Report of the Senate Committee on Education and Labor (LaFollette Report), 1942, pp. 37-38.

7. <u>Ibid</u>, p. 41.

8. Craig, <u>op cit</u>, pp. 38-9.

9. New York Times, April 5, 1959.

10. Ernesto Galarza wrote several books on California farmworkers, including <u>Merchants of Labor</u> (Santa Barbara: McNally and Loftin, 1964) and <u>Farmworkers and Agribusiness in California, 1947-1960</u> (Notre Dame: University Press, 1977).

11. The early years of the UFW are documented in Ron Taylor <u>Chavez and the Farmworkers</u> (Boston: Beacon Press, 1975) and Stephen Sosnick, <u>Hired Hands: Seasonal Workers in American Agriculture</u> (Santa Barbara: McNally and Loftin 1978).

12. Sosnick <u>op cit</u>.

13. Colin Carter <u>et al.</u>, "Agricultural Labor Strikes and Farmers' Income," <u>Economic Inquiry</u>, Vol. 25, No. 1, January 1987, pp. 121-133.

14. For example, the 12 employees of Riesner Nursery in Farmersville and the 40 workers at Baker Brothers citrus in Woodlake voted to decertify the UFW in

March-April 1987, but until the ALRB "certifies" these decertification votes, the UFW considers them "contracts."

15. <u>Christian Science Monitor</u>, June 22, 1987, p. 6.

Chapter 4

The ALRA and ALRB:
1975-1986

In 1975, the California Legislature adopted the Agricultural Labor Relations Act (ALRA) and created the Agricultural Labor Relations Board (ALRB) to administer it. The ALRA gives agricultural workers the right to ask the ALRB to supervise an election in which the workers choose a representative for the purpose of collective bargaining with their employer. The ALRA requires the union bargaining representative to be elected by majority vote in a secret ballot election. Unlike the National Labor Relations Act (NLRA), the ALRA does not permit an employer to voluntarily recognize a union as its bargaining representative.

The ALRA defines unfair labor practices which can be committed by an employer or by a labor union. The act authorizes the ALRB to conduct administrative hearings when workers, employers, or unions charge in written statements that unfair labor practices have been committed. The ALRB issues remedies when and if it agrees that someone has engaged in an unfair labor practice.

A. ELECTIONS AND CERTIFICATIONS

The union that wins a majority vote in the representation election becomes the *exclusive* bargaining agent for the workers. After a union wins the election, the workers must bargain with the employer through their union representative and the employer must bargain with the union, the workers' elected representative. Thus, the stakes in representation elections are high. Unions must request that an election be held, and unions do not request an election unless they are reasonably confident of winning a majority vote.

In the 12 years since the ALRA was enacted in 1975, the ALRB has supervised almost 1,100 elections on California farms. Unions were certified as bargaining representatives in two-thirds of these elections and the voting workers decided to have no union representative in 14 percent. About 95,000 valid farmworker votes were cast in these elections; although probably fewer than 95,000 workers voted because the same worker may vote on several farms. The work force represented by these votes is much larger than 95,000 because elections are sometimes held when only about one-half of a farm's peak work force is employed, and some eligible workers do not vote.

The ALRB must "certify" the results of each farm's election by deciding if the election was "fair" and then determine which union, if any, won; the ALRB can also "set aside" the election or refuse to certify it. The ALRB has certified about 850 of the 1,055 elections; 205 are either pending before the ALRB for certification or have been withdrawn or set aside, e.g., because the union or employer interfered to prevent a fair election. Of the 850 certified elections, workers on 382 farms (45 percent) selected the UFW as bargaining agent; 325 farms (38 percent) selected unions other than the UFW; and 144 farms (17 percent) selected no union. Most of the non-UFW union victories were at southern California dairies.

The goal of a union is to win an election and then negotiate a collective bargaining agreement. Not all certifications yield agreements. In 1984, about two-thirds of all certified elections yielded agreements in the year of the election, a proportion that has been decreasing.

The number of elections and certifications has been tabulated by year and region. This data must be interpreted carefully because the certification process can be lengthy. Workers, unions, or the employer can object to the outcome of an election by filing a technical objection or by alleging that misconduct occurred during the election campaign, such as threats or promises which affected the outcome of the election. The executive secretary of the ALRB reviews such election objections and orders a hearing to resolve meritorious objections—those which may have actually affected the outcome of the election. An investigative hearing examiner (IHE) presides over the election hearing and issues an opinion which recommends either that the election be set aside or rerun later or that the results are a valid indicator of worker preferences and the election should be certified. The IHE opinion can be and often is appealed to the five-member ALRB, which also makes a decision. Over the past 11 years objections were raised to about half of all elections (Table 4.1).

There have been two major periods of election activity: 1975-77 and 1980. Over 40 percent of all the elections that have been supervised by the ALRB occurred just after the ALRA was enacted in 1975. The number of elections fell sharply in 1976, in part because the ALRB ceased operations for six months. After elections were held on most of the farms that had been involved in pre-ALRA union activity, the number of elections fell to a rate of 30 to 40 each year.

Seven unions have been involved in ALRB-supervised elections, and the workers selected a union as their bargaining representative in 75 percent of the certified elections. Of the 650 certified elections won by unions, the UFW won 57

Table 4.1

ALRB Election Cases, Fiscal Years 1976-1987

	1975-1976	1976-1977	1977-1978	1978-1979	1979-1980	1980-1981	1981-1982	1982-1983	1983-1984	1984-1985	1985-1986	1986-1987	Total
Elections:													
Held	421	177	133	67	38	66	25	36	41	30	31	12	1077
Objected to	NA	—	—	35	17	43	14	21	26	18	13	7	194
Set for Hearing	—	—	—	29	9	23	14	12	8	6	9	5	115
Hearing:													
Opened	102	27	26	16	9	14	15	11	9	8	9	9	255
Decided	85	14	40	19	10	9	16	7	11	8	7	8	234
Board Decisions on													
Election Objections	65	30	26	31	5	4	17	9	11	4	5	4	211
Elections Set Aside	—	—	—	8	4	3	2	6	4	2	1	3	33
Elections Certified	171	158	210	72	42	34	44	30	38	23	13	17	852
Union Certification	165	152	190	59	30	28	38	17	21	12	4	5	721
No Union Certification	6	6	20	13	12	6	6	13	17	11	9	12	131

Source: *ALRB* Annual Reports.

Figures for FY 1986-87 are only through May 1987.

percent, the Christian Dairy Association won 33 percent; and other unions won 10 percent. Only half of these union victories yielded 1987 contracts, and the rate at which election victories are translated into union contracts varies across unions. Dairy workers unions have been most successful in translating election victories into contracts, and the UFW has been least successful—in 1987, the UFW had contracts on only one-sixth of the farms on which it had won elections.[1]

Farmworkers can also decertify the union which they elected to represent them and there have been 56 decertification elections involving 4,209 workers. In 27 of these, a union was decertified and 12 resulted in recertification of the UFW. The other decertification elections are still pending, were withdrawn, or were set aside by the ALRB.

B. UNFAIR LABOR PRACTICES

The ALRA, like the NLRA, created previously unrecognized rights for workers, including the right to self-organization, to form, join, or assist labor organizations; the right to bargain collectively through elected representatives; the right to engage in concerted activities; and the right to refrain from any of these activities.

These rights are enforced by a state-created agency, the ALRB, with a staff of about 200 and an annual budget of $8 million. Allegations that workers' rights have been violated are filed with regional offices of the ALRB as unfair labor practice charges against employers or unions. Charges are often filed during pre-election periods when a union or employer believes another party has interfered with the right of workers to make a free choice in the upcoming election. Unfair labor practices charges are also filed while the employer and union are bargaining for a contract and during the life of a collective bargaining agreement.[2]

A charge can be filed by any interested person against an employer, union, or an agent of either, and the ALRB investigates each charge to determine if it has merit. If the ALRB determines that a charge has merit, and if the charge cannot be settled through informal negotiation, the General Counsel of the ALRB issues a complaint containing one or more charges.[3] However, charges may be settled or withdrawn before a complaint is issued or the General Counsel may decline to issue a complaint, i.e., the General Counsel may decide that there was no violation and thus end the matter. If a complaint is issued, it is assigned a case number and set for a hearing before an administrative law judge (ALJ), who hears testimony from the charging and charged party and issues a decision. A complaint may be settled before, during, or after the hearing, and any party may file "exceptions" to the ALJ decision, or appeal the ALJ decision to the five member Board. Finally, either party may obtain review of a Board decision or order by a state court of appeal.

Table 4.2 traces unfair labor practice charges against employers or unions which were filed with the ALRB; Table 4.3 summarizes the disposition of these charges and complaints. Two trends are apparent: the number of charges filed has decreased, and the percentage of charges which resulted in complaints has dropped. In recent years, the number of charges dismissed by the General Counsel has increased, as has the percentage of charges which were settled by the parties.

Elections and unfair labor practice charges are the work of the ALRB, and trends in unfair labor practice charges figure prominently in the ongoing debate over whether the ALRB is "biased" in favor of growers or unions. Over 90 percent of all charges are filed against employers, so the high number of charges filed and the high percentage going to complaint between 1976 and 1983 fueled grower criticisms that the ALRB was "biased" against them. In the mid-1980s, the UFW has declared that the ALRB is biased against it because the General Counsel turns so few of the union's charges into complaints, so the UFW no longer files many charges with the ALRB.

The ongoing dispute between growers and unions over the "biases" of the ALRB makes it hard to predict trends in elections or unfair labor practice charges. However, it is clear that only a handful of farms have been involved in the decisions of the five member Board. Through 1987, the Board issued 760 decisions, but these decisions involved only 370 farm or union respondents. A few farm employers generated a disproportionate number of Board decisions, e.g. one farm was involved in 20 decisions, and fewer than 100 farms were involved in most of the Board decisions.

The decisions of the five member Board can be appealed to the California courts, and about 40 percent have been appealed. Such a high appeal rate is yet another indicator of the "immaturity" of farm labor relations—instead of accepting Board decisions, the losing party appeals to the courts. According to ALRB statistics, such appeals of Board decisions are rarely successful. However, "losing" parties continue to appeal Board decisions, suggesting that they have firmly-held convictions and highlighting the ambiguity of statistics on how often the Board's decisions are upheld by the California courts.

C. ORGANIZING AND COLLECTIVE BARGAINING

Unions must organize workers and win elections before they can bargain with the employer on behalf of the workers who elected them. The goal of a union is to obtain the best collective bargaining agreement possible. The goal of many employers is to operate without a union representative. Thus, each step on the road to a collective bargaining agreement—organizing, the election, and bargaining—is a high stakes campaign by both the union and the employer.

Table 4.2

Unfair Labor Practices, Fiscal Years 1976-1987

Items	1975-1976	1976-1977	1977-1978	1978-1979	1979-1980	1980-1981	1981-1982	1982-1983	1983-1984	1984-1985	1985-1986	1986-1987	Total
1. Charges Filed	926	767	737	814	1,302	938	930	1,234	885	732	453	234	9,952
2. Charges Withdrawn	90	99	170	120	279	160	195	164	102	58	7	21	1,465
3. Charges Dismissed	148	335	275	220	260	411	492	393	424	680	332	193	4,163
4. Charges Settled	9	22	35	9	16	6	12	43	81	59	28	27	347
5. Charges into Complaint	261	370	325	345	438	426	366	192	162	136	86	43	3,150
6. Charges Acted on (2+3+4+5)	508	826	805	694	993	1,003	1,065	792	769	933	453	284	9,125
7. Charges Added to Backlog (1-6)	418	-59	-68	120	309	-125	-135	442	116	-201	0	-50	767
8. Backlog	418	359	291	411	720	595	460	902	1,018	817	817	767	7,575
9. Complaints Issued	150	156	115	161	160	105	137	85	65	47	46	24	1,251
10. Complaints Withdrawn	3	14	7	6	16	10	13	14	11	15	7	4	120
11. Complaints Dismissed	2	4	4	3	11	3	1	2	2	0	0	1	33
12. Complaints Settled Prior to Hearing	10	35	36	23	15	22	28	16	22	8	28	7	250
13. Hearings Opened	52	039	97	78	94	87	90	110	45	35	24	18	869
14. Hearings Settled At or After Hearings	11	55	36	25	26	18	13	44	15	36	8	5	292
15. ALO Decisions	7	85	81	43	40	73	70	67	40	23	14	6	549
16. Board Decisions	1	23	53	71	48	35	62	83	29	31	32	22	510

Source: *ALRB* Annual Reports.
Figures for FY 86-87 are through May 1987.

Table 4.3

ALRB Disposition of Charges and Settlement of Complaints, Fiscal Years 1976-1987

	Fiscal Year											
	'76	'77	'78	'79	'80	'81	'82	'83	'84	'85	'86	'87
Percent of Charges:												
Withdrawn	9.7	12.9	23.1	14.7	21.4	17.1	21.0	13.3	11.5	7.9	1.6	9.0
Dismissed*	16.0	43.7	37.3	27.0	20.0	43.8	52.9	31.8	47.9	92.9	73.3	82.5
Settled*	1.0	2.9	4.7	1.1	1.2	0.6	1.3	3.5	9.2	8.0	6.2	11.5
To Complaint*	28.2	48.2	44.1	42.4	33.6	45.4	39.4	15.6	18.3	18.6	19.0	18.4
Acted On	54.9	107.7	109.2	85.2	76.2	106.9	114.6	64.2	86.9	127.4	100.0	121.4
Percent of Complaints Withdrawn	2.0	9.0	6.1	3.7	10.0	9.5	9.5	16.5	16.9	31.9	15.2	16.7
Percent of Complaints Settled:												
Before Hearing	6.7	22.4	31.3	14.3	9.4	21.0	20.4	18.8	33.8	17.0	60.9	29.2
At or After Hearing	7.3	35.3	31.3	15.5	16.3	17.1	9.5	51.2	23.1	76.6	17.4	20.8
Total Complaints Settled and Withdrawn	16.0	66.7	68.1	33.5	35.7	47.6	39.4	86.5	73.8	125.5	93.5	66.7

* Calculated as percent of charges filed in same fiscal year. Some of the charges which were dismissed or sent to complaint, however, may be from the backlog.

Source: Compiled from *ALRB* "Reports of Unfair Labor Practices."

Figures for FY 86-87 are through May 1987

Unions initiate most organizing campaigns, but there is an old union adage which asserts that management organizes workers into unions because of its mistreatment of employees, e.g. practicing favoritism in hiring or work assignments, paying substandard wages or benefits, or simply not being responsive to worker concerns. In most instances, a handful of aggrieved workers begin to voice their complaints to other employees, and they often conclude that "we ought to have a union." These workers sometimes contact the union, and sometimes a union organizer contacts them, but most union organizing campaigns are built around a handful of dissatisfied workers.

When the union formally enters the picture, a succession of papers must be filed with the ALRB. Unions first file a Notice of Intent to Obtain Access with the employer and the ALRB, and after an N/A is filed, the union may "take access" to the workers, i.e., union organizers may come onto the farm and talk to workers about the benefits of the union. This "access to private property by nonemployee union organizers" is controversial, because farmers who do not want to do anything to help the union to organize their workers allege that such access violates their private property rights.

The union organizer and the union's "inside employee supporters" attempt to get workers to show their support for the union by having them sign "authorization cards." These authorization cards simply state that the worker wants the union to represent him or her. The worker's signature on such a card is not binding under the ALRA, because the union must win an election before it can bargain for workers. However, the union, the employer, and the ALRB look at the percent of workers who have signed authorization cards as an indicator of whether there is enough union support to hold an election.

After 10 percent of the farm's employees have signed authorization cards, the union can file a Notice of Intent to Organize (N/O) with the ALRB, and then the employer must provide a list of all worker names and local residence addresses so the union can find and contact workers. As the organizing campaign continues, more workers sign cards, and when a majority of the workers have signed cards or an election petition, the union can request that the ALRB hold an election. Unions typically want 60 to 70 percent of the workers to sign cards supporting the union so that they can be assured of a majority vote in the election.

Since most farmworkers are employed for only two to six weeks on a farm, access, organizing, and elections occur within a very short period. There are charges and countercharges; the union wants to win an election, the employer wants to get the harvest completed and put off "labor problems" until next year. Unions and employers typically assert that the other party has violated the ALRA or election regulations by, for example, threatening to change crops if the union wins or calling out the Boarder Patrol to inspect worker documentation if the union is rejected. After the election is held and the votes counted, it is often hard to sort out the truth of these allegations and determine whether they affected the outcome of the election.

After the election is held and the votes are counted the ALRB must certify whether a union won the election. If the union is certified as the workers bargaining representative, then the employer and union are required to bargain "in good faith" to reach an agreement. Good faith bargaining has been likened to the ALRA requiring the union and the employer to enter a room and abide by the rules of the game, but then the ALRB must stand outside the room and decide on the basis of union and employer reports of what happened whether good faith bargaining occurred.

Bargaining sessions are usually straightforward. Most unions have a standard contract, and since the union initiated the drive to represent workers, the union typically asks the employer to sign its standard contract at the first session. The employer usually takes the union proposal, studies it, and then several weeks later makes a counterproposal. This proposal and counterproposal process continues until an agreement is reached, and then lawyers write up the final contract and the union and employer sign it.

Unions typically attempt to obtain a standard contract from all employers because they do not want an employer to say "you do not have that paid holiday at that farm, and I cannot afford it." Between 1975 and 1978, the UFW negotiated remarkably similar agreements, but after 1978, contract provisions began to vary as the UFW bargained for what it could get from the strongest employers and gave concessions to weaker employers.

A standard UFW contract includes clauses covering eight major subjects (Figure 4.1). It is similar to most nonfarm contracts in its union security and management rights clauses, except that the UFW contract requires farm workers to join the union after five days of employment (versus after 30 days in nonfarm industries) and it requires employees to remain union members "in good standing" to keep their jobs. Most UFW contracts establish a minimum hourly wage and then stipulate the rates to be paid to piece-rate workers, such as $13 for picking a bin of lemons or 90 cents for cutting a carton of lettuce. California farmworkers are not entitled by law to overtime pay until they work 10 hours daily, but most UFW contracts call for overtime after eight hours of work for all farmworkers except irrigators and tractor drivers.

Vacation, holiday and leave clauses reflect the peculiar nature of farmwork. Some vacation clauses provide more vacation pay to the employees who worked the most hours by having notches, e.g., 10 hours pay after 400 hours of work and 30 hours after 800 hours. Normal employee benefits include medical insurance, a pension plan, and employer payments for union-run social welfare funds. Unlike most nonfarm contracts, many UFW contracts include a housing clause which discusses charges for on-site housing and meals.

Figure 4.1

Major Provisions of Standard UFW Contracts

1. **Union Security and Management Rights**
   - Recognition of UFW as sole bargaining agent for farmworkers
   - Union shop (employees must join union after five days of work)
   - Dues checkoff (employer forwards 2 percent of worker pay to union)
   - Union health and safety committee to advise on work practices
   - Successor clause: agreement is binding on new owner
   - Management rights: all rights not stated in agreement are management rights
2. **Wages and Related Provisions**
   - Hourly or piece-rate wage structure and how it is to be adjusted
   - Piece-rate incentives and end-of-season bonuses
3. **Hours of Work and Overtime Pay**
   - Overtime pay after eight hours
   - Reporting pay e.g., four hours pay for each day the employee reports as instructed but there is no work
   - Premium pay for working at night
   - Travel pay for time spent riding to and from the worksite
   - Rest periods e.g., 15 minutes for every four hours worked
4. **Paid and Unpaid Leave**
   - Vacations: 2 to 4 percent of gross earnings
   - Paid Holidays: five to seven holidays for workers employed the day before and after the holiday
   - Leaves for union duty, jury duty, etc., without loss of seniority
5. **Employee Benefits**
   - RFK Medical Plan – employer contributes 75 cents to $1.25 for each hour worked
   - JDLC Pension Plan – employer contributes 15 cents to 20 cents for each hour worked
   - MLK Fund – employer contributes 5 cents to 15 cents for each hour worked
   - Jury, bereavement, and witness pay — employer pays at average rate
   - Citizen Participation Day — employer pays workers at their average rate for first Sunday in July
   - Housing: what access and cost if provided
6. **Seniority and Hiring**
   - Employer maintains a seniority list on the basis of hours worked
   - Union shall maintain a hiring hall to supply new workers
   - All new employees have a five-day probation
7. **Job Security**
   - Notify union of subcontracting necessary if workers do not have necessary equipment of skills
   - Mechanization: employer shall notify the union and bargain; union can call a strike if settlement is not reached.
8. **Dispute Settlement**
   - Grievance procedure: multi-step appeals to settle differences of opinion over what the contract means
   - There shall be no strikes and lockouts during the agreement

Source: Review of 250 UFW contracts

Rehiring and layoff by seniority is a basic principal of most union contracts. Because of the seasonal nature of the farmwork, most UFW contracts record seniority on an hours-worked basis. The UFW is concerned about farm jobs being eliminated by mechanization, so the UFW typically includes a clause that calls for bargaining over machines which will eliminate 25 percent or more of the farm's jobs. Finally, most UFW agreements call for a multi-step grievance procedure to resolve differences over what the contract really meant in a particular situation, e.g., if an employee damages a piece of machinery, is this grounds for dismissal or not?

This overview of how a union organizes, wins an election, and bargains for a contract has emphasized the speed with which the organizing and election process moves in seasonal agriculture and the complexity of contracts which must cover both working and living arrangements. After a first contract is signed, most nonfarm unions and employers "learn to live with each other," so that re-negotiation becomes a routine exercise emphasizing changes in wages and hourly contributions. However, the unstable nature of collective bargaining in agriculture has injected a considerable degree of uncertainty into re-negotiations; farms go out of business or change crops and reduce farmworker employment; disgruntled union members may seek a decertification vote; and both unions and employers seek to enlist the ALRB to support their charges and countercharges.

D. SUMMARY

The ALRA granted farmworkers the right form, join, or refrain from union activities, and to exercise these rights without interference from employers or unions. The seasonal nature of agriculture means that the ALRA provides for quick elections, and the ALRA also includes other features which affect bargaining and the relationship between a union and its members.

The ALRB supervises elections, and since 1975 the ALRB has supervised elections on about 1,100 California farms. On 650 farms, a union won the election, but these union elections translated into only 330 collective bargaining agreements in 1987.

The ALRB also considers charges that employers and unions have violated the ALRA by committing unfair labor practices (ULPs). About 90 percent of all ULPs, allege that the employer has violated the ALRA, and 60 to 80 percent of these ULP charges are withdrawn, dismissed, or settled by the ALRB.

An overview of union organizing and bargaining emphasizes the high stakes in the union's efforts to be the sole representative of farmworkers in all wage-related matters. Union organizers and inside employee supporters convince workers to vote for the union, a brief but intense campaign ensues, and then the union and employer are required to bargain in good faith to reach a collective bargaining agreement.

NOTES

1. Some of the farms on which the UFW won elections went out of business and on other farms, workers have voted to decertify the UFW.

2. Charges filed while the contract is in force may also be handled as grievances under the contract.

3. During bargaining, the parties frequently file charges against each other with the ALRB to gain leverage. Once an agreement is reached, these charges are usually dropped.

Chapter 5

Representation Elections

There are a number of differences between farm and nonfarm union elections, but the most important difference is the quick or expedited election under the ALRA. In California agriculture, elections must be held within seven days of a union's request for an election—and sometimes within 48 hours—so that the workers requesting an election on a farm have a chance to vote before the seasonal jobs on that farm end. The California Legislature believed that seasonal jobs and migrant farmworkers require quick elections in agriculture; Congress apparently felt that the more stable employment patterns of the nonfarm sector did not require that the NLRA specify an exact period during which an election must be held.

The other major differences between the ALRA and NLRA include the ALRA's wall-to-wall bargaining units, the ALRA's mandatory secret ballot election, and the ALRB's practice of waiting until after an election to resolve election-related issues. Wall-to-wall bargaining units mean that all the workers on a farm, both year-round and seasonal fieldworkers as well as clerks, and mechanics, pest advisors, and veterinarians must be included in one bargaining unit. This wall-to-wall bargaining unit requirement means that there can be only one union per farm, and the union must determine internally its priorities, such as e.g., balancing year-round worker demands for better pension benefits against seasonal worker demands for higher piece-rate wages.

Under the ALRA, the only way that a union can become the certified bargaining agent for farmworkers is to be selected by the majority of voters in a secret ballot election. This provision was included in the ALRA because the NLRA permits employers to voluntarily recognize a union as the bargaining agent for a group of workers, and in California agriculture during the early 1970s some growers switched from recognizing the UFW to accepting the Teamsters as the bargaining agent for farmworkers without elections or other evidence of worker desires.

The final difference between the ALRA and NLRA is that election-related issues are resolved after the election. For example, one ALRA election requirement is that at least 50 percent of a farm's peak number of workers must be employed at the time the election is held. Suppose the union asks the ALRB to supervise an election within seven days and asserts that employment is at least 50 percent of peak, and the union assertion appears to be true (e.g., the union files the petition in September). The employer may contest this union claim, saying that current employment is less than 50 percent of peak, but if an exact determination of peak employment is too complex to be made quickly, the ALRB will hold the election and then after the election determine whether employment was at least 50 percent of peak. Under the NLRA, in contrast, such technical objections to an election are resolved before the election is held.

A. ELECTION PROCEDURES

An election is a high-stakes gamble: if the union wins, it obtains the duty and obligation to represent all farmworkers on the farm in all matters related to wages and working conditions. If the employer "loses," it must bargain with the union over all wage and employment changes, that is, the employer cannot make *any* changes in wages or employment conditions without offering to bargain with the certified union representative. A union is normally certified as the bargaining agent for 12 months on a farm, so that even if the work force the next year is completely different from the work force which selected the union, the employer is obliged to bargain with the union representative. Selecting a union representative funnels all employer-worker negotiations through the union—the employer cannot make an independent "side deal" with a few favored workers.

The easiest way to understand how a union becomes the certified bargaining agent for the farmworkers on a farm is to follow the procedure from the union's filing of a petition asking that an election be held through the counting of ballots. Unions initiate the election procedure by filing a *petition* with the ALRB which requests that an election be held on a particular farm. The union election petition must assert that the farm named is an appropriate bargaining unit, that there is no certified union already representing workers, and that other technical requirements are satisfied. The union's petition must be signed by a majority of the workers on the employer's payroll or be accompanied by union authorization cards from a majority of the current employees (an authorization card is a card signed by a worker which gives a particular union authority to represent him or her). Unlike the NLRA, the ALRA specifically forbids employers from filing election petitions.

The union petition asserts that the workers on a particular farm are the appropriate *bargaining unit*, the group of workers for whom the union hopes to negotiate a contract. Unlike a retail store, which may have several bargaining units each with a different union, such as truck drivers represented by the Teamsters union, retail clerks by the Retail Clerks union, and butchers by the Butchers union, in California agriculture there can be only one wall-to-wall bargaining unit per farm. The ALRA included this wall-to-wall bargaining unit rule to prevent the fragmentation of collective bargaining; a farm employer must deal with only one farmworker union and not separate unions for each occupational group on the farm. Geographically separated operations of one farm will sometimes be treated as separate bargaining units.

The *timing* of a union election is critical, and the ALRA requires that union representation elections be held within seven days after a valid election petition is filed. A valid petition is one that is filed when current employment is at least 50 percent of peak employment and the union submits the required number of worker signatures.

The ALRB staff receives the petition and determines whether the union petition is "valid." This means that the ALRB checks with the employer to determine if it appears that current employment is at least 50 percent of peak employment; that a majority of workers signed authorization cards; that the workers' signatures are genuine; and that there are no *election bars* on a particular farm which prevent the holding of an election. Election bars include a currently certified union on a farm; a union contract which already covers the farmworkers (unless it expires within a year); or a valid election held on the farm within the previous 12 months.

The ALRB, the union, and the employer know that an election will be held seven days from the petition's filing date. At least 24 hours before the scheduled election day, union, ALRB, and employer representatives meet to discuss the exact time and place of the election (e.g., will voting be all day or just before and after work). These election details are decided in the *pre-election conference*, and then the ALRB prints and distributes notices in English and Spanish which inform workers of the details of the election.

Another union may *intervene* in the scheduled election, that is, have its name added to the ballot, if the second union can muster signatures from at least 20 percent of the employees at least 24 hours before the election. Most *elections are held* at or near the farm, usually in a shed, in a portable booth in the field, or in a labor camp, church, or public park. Ballots are printed in English, Spanish, and other appropriate languages, and the name and symbol of each union as well as the "no union" choice appear on the ballot. Workers usually line up to vote, mark their ballots in a closed booth, and drop them into sealed ballot boxes.

Workers approach a polling table staffed by the ALRB and a representative of the employer and the union(s). The union, employer, or the ALRB representative can "challenge" the right of any worker to vote by asserting e.g., that the worker who wants to vote was not on the payroll when the election petition was filed. Workers who are challenged vote *challenged ballots*; they vote just like other workers, but their ballots are separated. If the challenged ballots would not affect the outcome of the election, they are simply ignored (e.g., 10 challenged ballots in a 100 vote election in which workers voted 80 to 10 for a union will not be counted). If the number of challenged ballots could affect the outcome of the election (e.g., if the vote was 45 to 45) then each challenged ballot is resolved separately.

The *ballot count* is normally made as soon as the polls close. Ballot counts are public, so workers and union and employer representatives can watch the count. The workers whose ballots were challenged are investigated if the challenged ballots would affect the outcome of the election. When challenged ballots are counted, however, they are mixed together so that a worker's vote remains secret.

If a majority of the votes is not cast for a particular union or "no union," a runoff election is held within seven days. Voters eligible to vote in the first election are eligible to vote in the runoff election.

If a *strike* is in progress when the union files its election petition, the ALRB must hold the election within 48 hours. Strike election procedures are the same as nonstrike procedures with e.g., the pre-election conference, voter lists, and challenged ballots, but elections held during strikes are likely to include a polling place away from the farm.

B. POST-ELECTION ISSUES AND PROCEDURES

In nonfarm union elections, disputes about technical issues, such as the appropriate bargaining unit are settled *before* the election, but in agriculture such disputes are settled after the election has been held. After the ALRB announces the election outcome, the union and employer have five days to object to the election. These objections can allege that the election was flawed for technical reasons (*technical objections*), such as the union naming the wrong bargaining unit, or that the election was marred by improper *pre-election conduct* which prevented the workers from exercising their free choice. Unions and employers who object to elections are trying to get the election set aside or declared invalid.

The ALRB is reluctant to set aside elections because the seasonal nature of farm employment could require that the election be postponed for 12 months until employment is once again 50 percent of peak. Thus, the ALRB puts the burden of proof on the party which alleges that the election was invalid. If the ALRB agrees that the union or employer tried to coerce voters, then the Board must decide if this improper pre-election activity was sufficient to affect the outcome of the election. The investigative hearing examiner issues a decision and recommendation on the election in dispute, and either the union or employer can appeal this decision to the ALRB. The ALRB decision to certify a union cannot be appealed to the courts directly; however, if e.g. an employer refuses to bargain with the union because the employer believed that the election was invalid, the union can file an unfair labor practice charge. The employer's defense against the ULP charge is that the election and certification were invalid.

1. Technical Objections to an Election

There are four major issues that are disputed in farmworker union elections: who is the employer; was employment at least 50 percent of peak when the election was held; what is the appropriate bargaining unit on a farm with several operations; and exactly who is eligible to vote. The seasonal nature of agriculture, the migrancy of farmworkers, and the middlemen in many labor markets make these issues very difficult to resolve.

The ALRA's definition of an *agricultural employer*—"any person acting directly or indirectly in the interest of an employer in relation to an agricultural employee"—includes land owners and farm operators, custom harvesters, land management companies, and cooperative harvesting associations. The ALRA specifically excludes farm labor contractors (FLCs) from its definition of employers. The California Legislature believed that a stable bargaining relationship could not be established between a union and a farm labor contractor, since contractors often do not set wages or establish working conditions and some have no investments which would permit them to form stable employer-employee relationships. As a result, workers brought to a farm by a farm labor contractor are considered to be employees of the farm operator.

However, a person who supplies equipment in addition to workers and who is paid to harvest and transport the crop is considered to be a custom harvester and is the employer of the workers. It is often hard to determine if a particular "middleman" is a FLC or a custom harvester. The ALRB examines criteria such as the amount of managerial judgement or specialized equipment supplied by the FLC/ custom harvester.[1] Many farm employers want the entities who harvest crops to be considered custom harvesters so that these seasonal harvest workers cannot "vote in" a union and then move on. Consequently, in commodities such as citrus many "FLCs" now supply picking equipment and transport the fruit to a packing shed so that they are considered custom harvesters by the ALRB.

Two or more separate farms can be a joint employer. The ALRB has held that two companies are a joint employer when the operations of the companies are similar, there is employee interchange, a common labor relations policy, and identical ownership and management of the company. For example, in *Holtville Farms, Inc.* the ALRB found that common ownership, joint financial management, shared facilities, and centralized control of labor relations made two companies a single employer[2] (complete citations to cases are on the footnotes.) When some of these factors are the same and some differ between two companies, the ALRB examines the particular facts of the case to determine if the "joint employer" designation is warranted.

The seasonal nature of agriculture means that the number of workers on a farm fluctuates. On many farms, three or four workers are employed in September for each worker employed in February, and the workers employed in February are likely to be year-round clerks, mechanics, and livestock workers. However, all workers on a farm must be in one wall-to-wall bargaining unit, so the timing of an election will affect exactly who votes.

The ALRA states that an election can be held only when the total number of workers on an employer's payroll is *at least 50 percent of the peak* farm employment for the current calendar year. The germane payroll period is the one which precedes the union's election petition. Since most farm payroll periods are weekly, a petition filed on Monday would usually refer to the previous week. On many crop farms, this 50 percent of peak employment criterion limits elections to only three to six months each year.

Computing peak employment can be simple or complex. If all workers are on a weekly payroll, there is little turnover, and there is little seasonality, as in a dairy or livestock operation, then the ALRB simply checks the pre-petition payroll against the peak payroll and certifies that the 50 percent rule is satisfied. Complications arise when there are several payrolls or high turnover. For example, if field workers are paid weekly and office staff monthly, the ALRB has determined that employment shall be calculated by adding up both groups of workers and dividing by employment on a "representative" or average workday to compute the average daily number of workers (e.g., not a Sunday). If the peak employment period is two days, for example, but the payroll period is two weeks, a more complex calculation is required.

If there is high worker turnover, e.g., two workers in a crew of 30 quit and two are hired every day, then the ALRB must determine whether the average number of "job slots" (30 in this example) is at least 50 percent of peak employment. The ALRB compares the average number of workers employed each day during the week before the union petition was filed (30) to the average number employed each day during the peak period. If the peak period's average job slots are not more than 60 in this example, then a valid election can be held. This method of determining peak employment—by counting the average number of workers employed or "job slots"—is called the "Saikhon formula" because the board first announced the rule in *Saikhon* (1976) 2 ALRB No. 2.

A *bargaining unit* is a set of workers and jobs with a "community of interest" or similar circumstances and therefore likely to have similar representation desires. Unlike the NLRA, which permits the NLRB to decide the appropriate bargaining unit (whether e.g., the craft workers in an auto factory have a separate community of interest from assembly-line workers), the ALRA grants no discretion to the ALRB—the ALRA requires wall-to-wall bargaining units. The only exception to this wall-to-wall rule in agriculture is if the farm in question has two or more noncontiguous operations that do not share management, exchange workers, and satisfy other criteria.[3]

Unions sometimes petition for a single bargaining unit even if the employer's operations are 500 miles apart. The ALRB considers a variety of operational criteria to decide if the two operations should be a single bargaining unit (e.g., if a Salinas-based vegetable grower with operations in Imperial county is one or two bargaining units). These operational criteria include examining the centralization of management; the similarity of wages; the extent of common supervisory practices such as uniform work rules and employee handbooks; and any bargaining history.[4] Even if no workers are exchanged (e.g., because of fears that exchange workers may spread diseases among poultry), the ALRB has found that two poultry farms 90 miles apart are one bargaining unit because they share a centralized management and similar job structures.[5]

The fourth potential technical objection to an election outcome is a dispute about worker eligibility to vote. The ALRA stipulates that all agricultural employees who worked during the payroll period preceding the election petition are eligible to vote, including persons on paid sick leave, on paid vacations, or persons who worked but whose names did not appear on the payroll (all members of a family paid under one name are eligible to vote). There are no age or legal requirements that must be satisfied to vote in ALRB elections; both children and illegal aliens can vote.

Workers on strike are also entitled to vote. There are two kinds of strikes: unfair labor practice strikes called to protest an employer's unlawful activity (e.g., firing a worker for circulating a union election petition) and economic strikes (e.g. workers walk off the job to put pressures on the employer to offer higher wages). Unfair labor practice (ULP) strikers are eligible to vote, as are economic strikers who have been on strike less than 12 months.

Supervisory and managerial employees, guards, and close family relatives of the employer are ineligible to vote. Some farms also include nonfarm operations, such as a nonfarm packingshed or transportation enterprise, and such "nonfarm workers" cannot vote in ALRA elections. More complex are cases of "mixed work" employees—persons employed in both the farm and nonfarm operations of an employer. The ALRB has ruled that if packingshed workers also work regularly on the farm, they must be considered "farmworkers,"[6] but clerical employees are farmworkers only if the "bulk" of their work involves the farm operation.[7] All persons who appear at the polling place are allowed to vote, but if there is any doubt about their status or identity, they must cast challenged ballots.

2. Pre-Election Conduct

After the ALRB announces election results, either the union or the employer can object by alleging that the other party campaigned unlawfully or improperly, so that workers could not make a free choice in the polling booth. The NLRB has dealt with the question of proper campaign tactics by establishing a "laboratory condition" standard against which to judge campaign conduct—the NLRB has said that a union election is analogous to a laboratory experiment conducted under ideal conditions.[8] Anything which interferes with this laboratory condition can nullify the election, such as an employer or union *threat* of reprisals, e.g., such as the employer threatening to fire workers if they vote for the union or the union threatening to impose extra high initiation fees. Also interfering with worker free choice are *interrogations* of workers about their union activities, *promises* of benefits for voting in a certain way, or *surveillance* and other activities which create an atmosphere of fear. Under the NLRA, such TIPS (Threats, Interrogations, Promises, Surveillance) activities during the campaign require a new election and often result in the filing of an unfair labor practice charge against the offending party. If the NLRB finds that the employer's campaign tactics were especially egregious, it can order the employer to recognize and bargain with the union even if the union did not get a majority of the votes cast.[9]

The ALRB rejected the laboratory condition standard to judge campaign conduct; instead, the ALRB seeks to assure that farmworkers could make a "free and uncoerced choice" in the polling booth. This lower ALRB election standard means that unions and employers can commit more substantial TIPS campaign violations before the ALRB will order a new election. The ALRB examines such TIPS violations to determine whether they would have affected the outcome of the election. Under this less stringent ALRB campaign standard, if a union organizer threatened three workers, the election might not be overturned if other employees did not learn of the threats before the election, and the union won by a large margin. Similarly, the ALRB is reluctant to overturn elections if union (or employer) representatives improperly campaigned in the polling booth area unless the employer (or union) can prove that such campaigning clearly affected the outcome of the election.[10]

The ALRB and NLRB apply different standards to union agents than they do to union "adherents" (company employees who support the union).[11] The ALRB reasoned that conduct by union adherents which cannot be attributed to the union has less impact on the election and is less likely to effect the outcome of the election than similar conduct by nonemployee union organizers.[12]

The ALRB is reluctant to set aside or require new elections because the workers may have to wait another 12 months before an election can be held (until the 50 percent of peak employment criterion is satisfied). For example, if employer campaign misconduct prevented a fair election, the ALRB has worried that waiting

12 months for another election "rewards" the employer for improper activities, since it postpones for one year the obligation to bargain. However, the ALRB finds some types of pre-election conduct so coercive that a new election is required: threats of violence directed at workers, such as employer or union threats to call the Border Patrol to round-up illegal alien workers; increases in wages or benefits right before the election; or surveillance, such as when a supervisor read the names of UFW sympathizers to the entire crew before the election.[13] In an extreme case, the ALRB has ordered an employer to bargain with a union which "lost" the election because the employer's TIPS conduct was especially egregious.[14]

ALRB decisions on TIPS campaign tactics illustrate the real-life difficulty of conducting fair union elections. The stakes are high, and few elections are totally "clean." After the election has been held, ALRB staff must review what actually happened and then decide if the campaign tactics actually affected the outcome of the election. As might be imagined, partisan employer and union representatives have different opinions on how a particular event influenced workers, making the resolution of these TIPS campaign issues very difficult.

Resolving election issues often delays collective bargaining, sometimes generating worker resentment against the union. For example, workers at Cattle Valley Farms in Coachella selected the UFW as their bargaining agent in 1978, but the first contract was not negotiated until 1981. Between 1978 and 1981, many of the workers who voted for the union left, and some of the 1981 workers tried to decertify the UFW when informed that a contract had been signed. There was also a six year delay between the election and the first contract at Santa Clara Produce in Oxnard; this delay considerably cooled the union supporters' ardor and once again the contract applied largely to workers who had not voted for the union.

Farm and nonfarm union election procedures are compared in Figure 5.1 The major differences are the expedited elections in agriculture, the post-election resolution of disputes, and wall-to-wall bargaining units.

C. ACCESS

The ALRA grants farmworkers the right to organize for the purpose of collective bargaining, but in order to decide whether they want a union to represent them, workers must know what the union will be able to do for them. The burden of informing workers about the benefits of union representation falls on the union; a task made much easier if one or more employees is pro-union. These worker-organizers can praise the union and solicit authorization card signatures without employer interference. More complicated is the issue of access by nonemployee union organizers to workers while they are on the employer's farm.

Unions usually organize workers by sending organizers from farm to farm to extol the virtues of the union. Employers who do not want workers to be represented

Figure 5.1

ALRB and NLRB Election Procedures

Procedure	ALRB	NLRB
Petition may be filed by:	Employee Group of employees Labor organization	Employee Group of employees Labor organization Employer
Showing of interest required (for petitioner):	50 percent	30 percent
Time allotted for investigation of petition:	7 days	Unlimited, generally completed within 30 days.
Hearing on election issues:	Post-election	Pre-election and post-election.
Who can collectively bargain with employer:	Labor organization certified after a secret ballot election.	Labor organization certified after a secret ballot election. Labor organization recognized by employer.
Bargaining unit:	All agricultural employees; noncontiguous units may be combined or not.	NLRB determines appropriate unit; can be employer unit, craft unit, plant unit, or a subdivision.
Contract bar:	Up to 3 years, except that a petition for election can be filed during the last year of a contract.	Up to 3 years except that a petition for an election can be filed not more than 90 days but over 60 days before the contract expires.
Holding an election	By board order only.	By board order or by agreement between the parties.
Eligibility period for voting:	Pay period preceding filing of the petition.	Voters must be employed during eligibility period set by the Board and on the date of the election.

by a union are naturally reluctant to permit nonemployee union organizers to come onto their property to educate workers about the virtues of the union. ALRB *access rules* give nonemployee union organizers a limited right to enter a farm and talk to workers about their collective bargaining rights.

The ALRB examined the farm labor market and concluded that many workers live and work on employer property; that workers who enter and leave a farm come and go at different points, so that there is no single public place for organizers to talk to workers; and that worker characteristics limit the ability of organizers to communicate with workers through printed literature, radio, or other media. These farm labor features and the quick elections mandated by the ALRA prompted the ALRB to issue access regulations that allow union organizers to discuss the ALRA with workers on employer property. Union organizers have limited access rights: no more than two organizers for the first 30 workers and then one organizer for each additional 15, and no more than three hours of access per day—one hour before work, during lunch, and after work. In order to obtain access, the union must file with the ALRB and the employer a Notice of Intent to Obtain Access (N/A); by filing N/As, union organizers have access for up to four 30 day periods per year.

The N/A permits union organizers to come into the employer's property, but it does not tell the union exactly how many and where workers are working on the farm. As soon as the union has obtained authorization signatures from at least 10 percent of the employees in the bargaining unit, the union may file with the ALRB a Notice of Intent to Organize (N/O). The N/O requires employers to provide the ALRB with a complete list of worker names, street addresses, and job classification within five days. The ALRB then gives this employee list to the union.

If an election is held, unions have access for up to five days after the ballots are counted or up to 10 days after objections to the election are filed. After the union is certified by the ALRB as the bargaining agent for workers on a farm, the union has continued "post-certification access" to determine what issues workers want the union to negotiate with the employer. If the certified union calls a strike to put pressure on the employer during contract negotiations, the union retains "strike access" to workers who continue to work.

The access rule has been very controversial. When the access rule was adopted by the ALRB (it is not in the ALRA), farm employers alleged that the ALRB was "biased" again them. Indeed, employer opposition to the access rule was a major reason why the ALRB was without funds for part of 1976, and the growers' successful campaign against the UFW-sponsored Proposition 14 in 1976 hinged largely on appeals to voters that the access rule was a violation of private property rights.

Employers have long contended that it is unconstitutional to require them to permit nonemployee organizers to come onto their property. In 1943, the Supreme Court ruled that employers may not restrict the right of employees to organize fellow employees, and later the Supreme Court ruled that nonemployee union organizers must be permitted to talk to workers on employer property if the union has no alternative means of communicating with workers. The NLRB thus decides whether a union should have access to workers on employer property on a case-by-case basis, i.e., the union asks for access and explains why it is necessary. The ALRB, in contrast, reviewed the farm labor market and concluded that it would normally be asked to and then would grant access to nonemployee organizers, so it handled the issue by automatically granting limited access. After much litigation, the ALRB access rule was upheld by the U.S. Supreme Court.[15]

The ALRB enforces its access regulations in three types of proceedings: election objections, unfair labor practices, and motions to deny access. A union can object to an election when an employer violates the access rule by denying access to the union or by discriminating against one union in favor of another when granting access. If the ALRB finds that employer or union violations of the access regulation could have affected the outcome of the election, it can set aside the election.

If access violations are found to have coerced or restrained Wemployees, the ALRB can rule that an unfair labor practice was committed. In *Western Tomato Growers & Shippers, Inc.*, the ALRB found that employer representatives committed an unfair labor practice by carrying firearms, threatening union organizers and denying organizers access to the fields to talk to workers.[16]

When union representatives significantly or repeatedly violate the access rule, an employer can file a motion to deny access. The remedy is an ALRB order banning the named organizer(s) from taking access to any farm in the region for a designated time period. The ALRB grants a motion to deny access when violations of the access rule cause a significant disruption of agricultural operations; when there is intentional harassment of an employer or employees; or if there is intentional or reckless disregard of the access rule. In *Ranch No. 1, Inc.*, the ALRB held that a UFW organizer intentionally disregarded the access rule when he remained in a field one and one-half to two hours instead of the one hour permitted and his presence resulted in disruption of agricultural operations. This organizer was banned from all farms in the region for 60 days.[17]

ALRB access rules do not apply to employer-owned labor camps. Union agents are entitled to unrestricted access to on-farm housing, subject only to the owner's reasonable regulations to prevent interference with business operations.[18]

D. SUCCESSORSHIP

One objective of national labor policy, as reflected in the NLRA, is that employees who have elected a representative for collective bargaining purposes should be protected from sudden changes in wages and working conditions due to a change in the ownership of the company. Owners can buy or sell their businesses, but owner rights to sell their property must be balanced against protection for employees who voted for a union to represent them or signed a multi-year contract with the owner. *Successorship* rules determine when a new employer must abide by the election results or contract agreement of the previous employer. In the nonfarm economy, the labor obligations of the buyer are determined on a case-by-case analysis of factors such as continuity of the work force and continuity of the operations.

The ALRB also has a successorship doctrine. However, the ALRB has determined that the seasonal nature of agricultural employment, migration, the nature of the work, and the use of farm labor contractors all contribute to high worker turnover. Therefore, the ALRB has ruled that continuity of the workforce will not be the determining factor in successorship in agriculture. In a case involving a vegetable farm, the buyer became the successor employer to the collective bargaining obligation of the seller because it farmed the same land, used the same equipment and handled the crops in essentially the same manner. The size and the nature of the bargaining unit did not change, so the new owner had an obligation to the UFW (the worker-elected bargaining representative) even though the work force had changed.[19]

Once the ALRB determines that the buyer is a successor employer, then the buyer has a duty to bargain in good faith with the certified bargaining representative of the employees. The buyer will also be held liable for unfair labor practices of the predecessor if the buyer knew that the unfair labor practice dispute existed.

E. DECERTIFICATION AND RIVAL UNION ELECTIONS

Under the ALRA, employees can initiate an election to decertify a union by filing a petition (1) signed by at least 30 percent of the agricultural employees in the bargaining unit (2) during the last year of the contract (3) when 50 percent of the peak work force is employed. Two of the same election bars apply as in a representation election: a valid election cannot have been held within the preceding 12 months and the incumbent union cannot have been certified within the preceding 12 months.

A decertification election must be held in the same bargaining unit as the certification election. Procedures for holding a decertification election—such as voter eligibility and TIPS campaign restrictions—are the same as for a representation election. In a decertification election, an incumbent union must win a majority of votes to maintain its certification.

A rival union can attempt to be elected in a certified bargaining unit during the last year of a collective bargaining agreement by filing a petition signed by a majority of the work force. As in other elections, at least 50 percent of the peak work force must be on the eligibility list and there can have been no election or certification within the last 12 months. If a contract has expired, then a decertification election or rival union election can be held using the standard representation procedures, as long as there are no election bars.

Decertification of a bargaining representative is a remedy only for employees. Because employer assistance of a decertification effort interferes with employee free choice, employers are prohibited from filing decertification petitions and from encouraging decertifications by, for example, providing leaves of absences or other benefits to employees who organize the effort, or by arranging free legal advice for the employees. An employer can, however, respond to employees' questions concerning their rights to decertify the union and then refer them to someone for assistance.

Sometimes a union has not been decertified, but an employer refuses to bargain with the incumbent union because it believes that the union no longer has majority support among the workers. The NLRB and ALRB have held that the employer may be committing an unfair labor practice by refusing to bargain. Until the results of a decertification election have been certified by the board, the union remains the bargaining representative of the employees.

F. SUMMARY

A union can become the certified collective bargaining agent for farmworkers in California only by winning a certification election. Elections under the ALRA differ from nonfarm elections in three major ways; they are held in an expedited manner, they involve wall-to-wall bargaining units, and election issues are decided after the election. Representation elections can only be held when the employer's work force is at least 50 percent of peak employment, after a union has collected signatures from a majority of the workers, and when there are no election bars. Elections are conducted by ALRB staff within seven days of the filing of a valid election petition.

After the election, either party can raise technical objections to the election (allegations that it was invalid) or to allege that improper pre-election conduct prevented a fair election. The ALRB is reluctant to set aside elections because another election might not be possible for another year, so the Board asks whether or not improper conduct could have affected the election results ALRB access regulations guarantee union organizers access to company property before and after an election. Once a union is certified as the bargaining representative on a farm, the employer must bargain with the union. This duty to bargain sometimes transfers to the new owner if the farm is sold, according to successorship principles, and continues until the union is decertified in a decertification election or unless the employer goes out of business.

NOTES

1. Sutti Farms (1980) 6 ALRB No.11.
2. (1984) 10 ALRB No. 49.
3. John Elmore Farms (1977) 3 ALRB No. 16.
4. Bruce Church (1976) 2 ALRB No. 38.
5. Prohoroff Poultry Farms (1983) 9 ALRB No. 68.
6. R. C. Walters and Sons (1976) 2 ALRB No. 14.
7. Joe Maggio Inc. (1979) 5 ALRB No. 26.
8. General Shoe Corp. (1948) 77 NLRB No. 124 at 127.
9. Gissel Packing Co. v. NLRB, 395 U.S. 575 (1969).
10. D'Arrigo Bros. of California (1977) 3 ALRB No. 37.
11. Muranaka Farms (1986) 12 ALRB No. 9; NLRB v. Southern Metal Service 606 F. 2d 512 (5th Cir. 1979).
12. Muranaka Farms (1986) 12 ALRB No. 9 at 6.
13. Royal Packing Co. (1979) 5 ALRB No. 31, rev'd on other grounds 101 Cal. App. 3d 826 (4th dist. 1980).
14. Harry Carian Sales (1980) 6 ALRB No. 55, enforced, 39 Cal. 3d 209, 216.
15. Kubo and Pandol v. ALRB, 429 U.S. 802 (1976) (Hg. den.).
16. (1977) 3 ALRB No. 51.
17. (1979) 5 ALRB No. 36.
18. United Farm Workers of America v. ALRB (Sam Andrews' Sons), 190 Cal. App. 3d. 1467 (2d dist. 1987).
19. Highland Ranch and San Clemente Ranch (1979) 5 ALRB No. 54, modified and remanded, 29 Cal. 3d 874 (1981), modified, (1982) 8 ALRB No. 11.

Chapter 6

Unfair Labor Practices

The ALRA, like the NLRA, grants rights to workers and is primarily concerned with ensuring that workers' rights are not violated. The ALRA, for example, states:

> It is hereby stated to be the policy of the State of California to encourage and protect the right of agricultural employees to full freedom of association, self-organization, and designation of representatives of their own choosing, to negotiate the terms and conditions of their employment...[1]

The California Legislature hoped that granting organizing and bargaining rights to agricultural workers would ensure peace in the fields and bring certainty and a sense of fair play to agricultural labor relations.[2]

The specific rights of workers are established in Section 1152 of the ALRA.

> Employees shall have the right to self-organization, to form, join or assist labor organizations, to bargain collectively through representatives of their own choosing, and to engage in other concerted activities for the purpose of collective bargaining or other mutual aid or protection, and shall also have the right to refrain from any or all such activities...

The ALRA protects these worker rights by prohibiting certain union and employer conduct. Employer and union activities which violate these rights are called unfair labor practices (ULPs). The ALRA does not list all of the specific acts that are unlawful; it only establishes broad guidelines. Workers, unions, and employers file charges alleging that ULPs have been committed. The ALRB then decides whether these charges have merit, e.g. if a union-supporter is fired for drinking on the job, is the firing a lawful employer act or a ULP meant to discourage union activity. Unfair labor practices prohibited by the ALRA closely parallel those in the NLRA, and the ALRB must rely on NLRB decisions where they are applicable.

A. EMPLOYER UNFAIR LABOR PRACTICES

1. Section 1153(a): Interference, Restraint, or Coercion

Unfair labor practices by an employer are enumerated in section 1153 of the ALRA. Section 1153(a) prohibits an agricultural employer from interfering with, restraining, or coercing agricultural employees in the exercise of their rights guaranteed by section 1152. Violations of sections 1153(b), (c), (d), and (e) are also considered violations of 1153(a), since they also involve coercion or interference, so that almost all ULP charges against employers include the allegation that an 1153(a) ULP was committed. However, some employer conduct violates section 1153(a) alone, including employer interference with the right of self-organization and employer interference with the right to engage in concerted activities.

a. Interference with the Right of Self-Organization

The essence of the right of self-organization means preserving an employee's "freedom of choice" to select a bargaining representative without employer interference, restraint, or coercion. The ALRA leaves to the ALRB the task of deciding what conduct constitutes interference and NLRB and ALRB precedent have established certain types of conduct which violate section 1153(a).

Violence and physical intimidation of employees are obvious forms of employer interference. Violence against union organizers also intimidates workers and is prohibited. In *Perry Farms, Inc.*, the Board held that Ernest Perry violated section 1153(a) by shouting at organizers, pushing and shoving them and precipitating a fight in which an organizer's mustache was partially pulled off.[3]

Threats of violence to workers or in the presence of workers also violate section 1153(a). For example, preventing union organizers from talking to workers by brandishing firearms and threatening bodily harm was found to be an unfair labor practice in *Western Tomato Growers & Shippers, Inc.*[4]

Threats of economic reprisals interfere with workers' free choice. For many years the NLRB has ruled that certain statements by an employer, supervisor, other agent of the employer are violations, such as a statement that the employer will go out of business or will move the business to another area if the union wins an election. A statement that a farmer will substitute alfalfa for strawberries if the union wins an election is considered a threat of economic reprisal and a violation of section 1153(a), since alfalfa is not a labor intensive crop and the statement implies layoffs. Similarly, statements to workers that they will have no work if they sign union authorization cards violate section 1153(a).[5]

Granting wage increases or benefits, or promising benefits during an organizational campaign interferes with protected rights by using allurement instead of coercion. Since employers provide wage and benefit increases, an employer promise of such an increase may imply to workers that employers will not grant future increases if the workers vote for the union. In *Anderson Farms Co.*,[6] the ALRB held that announcing new benefits shortly before an election, even though the benefits had gone into effect several months earlier, was a coercive use of the employer's economic power and a violation of section 1153(a).

Surveillance of employees' union activities or giving the impression of surveillance is considered an infringement on the free exercise of employee rights. Section 1153(a) violations found by the ALRB include programs of monitoring and photographing union organizers at the work place and at workers' homes[7] and more subtle surveillance by a ranch manager who sat in his pickup truck holding paper and pencil while watching a union organizer talking to workers.[8]

Interrogation or questioning of an employee about his or her union views, sympathies, or activities tends to restrain and interfere with an employee's exercise of section 1152 rights. In *Rod McLellan Company* the ALRB held that asking an employee about her views, sympathies or activities or those of her husband and fellow workers violated section 1153(a).[9] Even an amicable conversation can be an unfair labor practice[10] if it involves questions about an employee's union sympathies, or questioning workers about their conversations with a union organizer.[11]

Denying union organizers access to the employer's premises, as required by the access rule, or to labor camps interferes with employee rights to self-organization,[12] as does denying organizers the opportunity to distribute literature. The ALRB has determined that the distribution of literature furthers the goal of effectively informing employees about the issues involved in unionization and is permitted by the access rule as long as distribution doesn't disrupt the employers' business.[13] Similarly, refusal to supply a complete and accurate employee list when required[14] interferes with employee rights because of the crucial importance of such lists to the employees' right to receive information.[15]

b. *Interference with Concerted Activities*

Employers can also violate worker rights by interfering with the activities of two or more workers who are engaged in activities for "mutual aid or protection." It is unlawful for an employer to interfere with such concerted activities, even if no union is involved. Employer interference, restraint, or coercion of workers engaged in concerted activity is a violation of section 1153(a).

Concerted activities usually involve two or more employees acting on behalf of a group of employees. However, complaints by a single employee to state officials about safety conditions on the job have been held to be protected by section 1153(a), since the complaints related to conditions of employment which concerned all employees.[16]

Protected concerted activities include a strike, walkout or work stoppage growing out of a labor dispute concerning terms, tenure, or conditions of employment. In *NLRB v. Washington Aluminum Co.*[17] seven employees left work just as the workday started because the furnace in the machine shop was not working and the outside temperature was below 22 degrees. The U.S. Supreme Court upheld the NLRB finding that the conduct of the workers was a concerted activity to protest the company's failure to supply adequate heat and that it was protected under section 7 of the NLRA (equivalent to ALRA section 1152). The workers were protected by Section 7 and could not be fired simply because they failed to make a specific demand for heat before walking out. Having no bargaining representative or negotiating procedure, they took the most direct course of action to let the company know they wanted adequate heat. The company, however, fired the workers for walking out and defended the firings on the basis of a plant rule which forbade employees from leaving their work stations without permission of their foreman. The Court acknowledged that the NLRA authorizes an employer to discharge employees for "cause," but does not allow an employer to punish an employee for engaging in protected concerted activities.[18]

In a similar case,[19] the ALRB found that an employer unlawfully discharged a crew of workers for engaging in a walkout for a number of reasons including extreme heat (103 degrees), exhaustion, lack of cool drinking water, and the mistaken belief that they were being sent to re-pick a melon field that would generate low piece-rate wages. Although a series of intermittent strikes is prohibited by both the NLRB and ALRB, a single concerted refusal to work is a protected activity, provided it is not violent, unlawful, in breach of contract (a no-strike clause, for example), or "indefensibly disloyal."[20]

2. Section 1153(b): Company Domination or Assistance of a Labor Organization

Section 1153(b) of the ALRA prohibits an employer from dominating, interfering with, or giving financial or other support to any labor organization. The same provision was included in the NLRA in 1935 to deal with the large number of company unions employers had created and controlled. Testimony before Congress indicated that if employers and unions competed to organize workers, bitterness and strife in labor relations often increased,[21] and that a company union whose representatives are selected or supported by the employer cannot command the full confidence of employees.

Employer interference which constitutes company domination and assistance of a labor organization includes direct financial support to selected workers or a particular union, participation in drafting the union constitution or bylaws, participation in internal union management or elections, supervising union meetings, paying employees for attending meetings and a wide variety of other practices. However, a violation occurs only if the employer's involvement is so great as to infringe on the exercise of employees' free choice.[22] If the ALRB finds that the employer dominated the organization by playing a major part in its formation or operation, the ALRB orders the disestablishment of the union.[23]

In many ALRB cases of unlawful employer assistance, the violation involves unequal treatment of two unions or granting one union more or less access than is granted to the other. In *Jack G. Zaninovich*[24] access was granted to one union, but denied to another. In *Royal Packing Co.*[25] both unions were granted access, but one union was given additional opportunities for access and assistance in soliciting authorization cards. In *E. & J. Gallo Winery*[26] the Teamsters were allowed access without interference while UFW representatives were followed and photographed whenever they attempted to communicate with workers. The company also campaigned on behalf of the Teamsters and condoned or assisted coercive actions by Teamsters against employees who supported the UFW. If an employer claims that additional access must be given to one union because of a contract with that union, the ALRB has held that the employer has the burden of proving that the favored union representatives limit their activities to legitimate union business, which does not include organizing or campaigning.[27]

New forms of worker participation in management, such as quality circles, have presented the NLRB and federal courts with additional complexity in interpreting the prohibition against employer domination and assistance of a union. Quality circles involve some form of employer-employee cooperation in certain decision-making functions. Common in Europe, such cooperation is less frequent in the United States. Under a Supreme Court decision, employee committees to discuss problems of mutual interest to workers and employers and to handle grievances at nonunion plants were found to be unlawful because of the prohibition against employer dominance or assistance,[28] especially if the employer retained the power to review or overrule committee decisions. A more recent court of appeals decision, however, indicates the court's willingness to consider cooperative arrangements lawful. In *Hertzka and Knowles v. NLRB*,[29] the court found that a committee system to deal with employer-employee relations, which was initiated on an employee's suggestion and approved by the employees, did not violate the NLRB despite the fact that meetings were held on company time and management partners participated in committee meetings. The court concluded that the line between lawful employer-employee cooperation, encouraged by the Act, and actual employer interference or domination depends on whether the employees' free choice is stifled by the employer involvement.

3. Section 1153(c): Employer Discrimination in Conditions of Employment

Section 1153(c) prohibits an employer from discriminating in hiring or firing, or any other condition of employment, in order to encourage or discourage membership in a labor organization. Most unfair labor practice cases before the ALRB and NLRB involve employer discrimination against workers who are sympathetic to unions. An aggrieved worker must prove four elements in a discrimination case against an employer: (1) that there was a discriminatory act by the employer against one or more employees; (2) that the worker was engaged in union activities; (3) that the employer had knowledge of the worker's union activities; and (4) that the employer had an anti-union motivation for discriminating against the worker. The employer's motivation or state of mind is the most difficult element to prove and must often be inferred from the employer's conduct.

a. The Discriminatory Act

An employer's discriminatory act may be discharge, layoff, refusal to hire, rehire, or recall, or constructive discharge (forcing a worker to quit). In *Kaplan Fruit and Produce Co.*,[30] for example, the employer violated section 1153(c) by discharging a crew boss and his crew to retaliate against the crew's union activities. In *Pioneer Nursery*[31] supervisors segregated pro-union and anti-union employees into different crews and threatened to offer less work if the union won the election. Immediately after the union won the election, pro-union employees were laid off.

Discharge of a supervisor in retaliation for his wife's union activities was found unlawful in *McAnally Enterprises*.[32] In *O. P. Murphy Produce Company*[33] the employer refused to rehire all members of several families because of union activity by some family members. Refusal to rehire undocumented workers who were union activists was found unlawful in *Nishi Greenhouse*.[34] The layoff of a union supporter by *Sam Andrews' Sons*[35] was found not to be discriminatory, but failure to rehire him when work increased later in the season was found to be discriminatory.

Discrimination against union activists in working conditions is also an employer violation of 1153(c). Such violations include demoting union supporters or assigning them more difficult, onerous or dangerous tasks; transferring workers to less desirable jobs or a less desirable location; and discriminatory enforcement of work rules. For example, the transfer of two employees from their normal work to pulling noxious weeds, which injured their hands and which precipitated their quitting, was found to be an unlawful constructive discharge because it was motivated by their union activities.[36] Eviction from company housing in retaliation for union activity was found unlawful in *Filice Estate Vineyards*.[37] In *Mike Yurosek & Son, Inc*.[38] the employer unlawfully assigned a union spokesperson permanently to a position that had previously been alternated among all crew members and was more dangerous. In *Paul Bertuccio*[39] the employer unlawfully assigned more difficult work in an uncultivated field to a pro-union crew.

Often an employer accused of unlawful discrimination presents a plausible reason for the discriminatory treatment of a worker, such as firing a pro-union worker because he was late consistently or because he was drinking on the job. In such cases where the employer has both a lawful reason for firing a worker (being late or drinking) and an unlawful reason (union activities), the worker must prove that there was a discriminatory act by the employer; that the worker was engaged in union activities; that the employer knew of these union activities; and that *one* reason for the discrimination (firing) was the employer's anti-union motivation. After the worker establishes such a case, the employer must prove that the same policy would have been followed if there had been no union activity, e.g. all workers who are late or drink on the job are fired.

In *Paul Bertuccio*[40] the employer claimed that he denied field work to five union activists because they had no experience in such work. His defense was discredited, however, because he hired other new workers who had no prior experience to do the same work. In another case,[41] the employer's defense that there was no work available when the discriminatees applied was discredited by a showing that he gave work to anti-union employees only. An employer's defense that he demoted a union supporter because of alleged misconduct was dismissed in *Rigi Agricultural Services, Inc*.[42] because the demotion was disproportionate to the misconduct.

b. *Responsibility for Discriminatory Acts*

Under section 1153(c), as in other unfair labor practices, an employer is responsible for the misconduct of his or her agents, including supervisors and others who represent the employer. Additionally, the employer is responsible for unfair labor practices committed by a farm labor contractor he or she employs,[43] since a farm labor contractor is not an employer under the ALRA.[44]

c. Motivation

A discriminatory act violates section 1153(c) only if the employer's motivation is to encourage or discourage union membership. Since an employer rarely expresses his or her reasons for the conduct, the Board must draw inferences from the evidence. An employer defense can sometimes be discredited by contradictory evidence, such as the hiring of nonunion employees after telling union supporters that there is no work available. The Board may infer that since the stated reason was untrue, the underlying motive was an unlawful one. In other cases, the surrounding events lead to an inference that the employer's motivation was anti-union. For example, in *Anderson Farms Company*[45] a supervisor fired six workers after observing them talking with UFW organizers; the ALRB considered the firings an anti-union threat and inferred that the employer had an anti-union motive.

When there are two or more motives advanced for the employer's discriminatory conduct, so-called dual motive cases, the worker and the ALRB General Counsel must show that the protected activity was *one* of the motivating factors in the employer's decision. The burden then shifts to the employer to show that he or she would have made the same decision even if the worker had not engaged in the protected union activity.[46] In *Nishi Greenhouse*[47] two workers were apprehended by the Immigration and Naturalization Service (INS). After returning to the United States, they asked for their old jobs back, but Nishi refused to rehire them. The two had been union supporters and had distributed union leaflets. The employer had engaged in anti-union tactics, such as soliciting grievances, granting unusual benefits and changing company rules to prevent workers from communicating about union matters. The ALRB General Counsel showed that Nishi had discriminated by not rehiring the two illegal aliens who were union supporters; the two had engaged in union activities; Nishi knew of their union activities; and one reason for refusing to rehire them was Nishi's anti-union motivation.

With the four necessary elements of section 1153(c) proven, the burden shifted to Nishi to prove that he would not have rehired the workers even if they had not engaged in union activities. Nishi claimed that he did not rehire the workers because four or five months earlier he had instituted a policy of not rehiring workers who had been apprehended by the INS, believing that he could be subject to a large fine for hiring undocumented workers. This asserted justification was belied, however, by the fact that Nishi did rehire five other workers who had been apprehended by the INS. The evidence also showed that Nishi did not put the new illegal alien policy in writing or inform his employees about it, nor did he take any steps to avoid hiring undocumented workers in the first place. The ALRB concluded that the illegal alien workers were not rehired because of their union activities in violation of section 1153(c). (The ALRA protects farmworkers regardless of their legal status, so that undocumented workers may vote in ALRB elections.)

In a few cases, an employer's discriminatory conduct is inherently destructive of employee rights and it is presumed to violate section 1153(c) even if there is no specific proof that he or she intended to discourage union membership. An example of such conduct occurred in *NLRB v. Great Dane Trailers, Inc.*[48] where the company announced a new policy that gave vacation pay to all employees who were at work on July 1. However, that date fell during a strike, so only nonstrikers were given vacation pay. The Supreme Court found that this conduct was inherently destructive of employees' rights to organize and therefore violated NLRA section 8(a)(3), the equivalent of ALRA section 1153(c).

4. Section 1153(d): Employer Discrimination for Participation in Board Proceedings

Section 1153(d) of the ALRA prohibits agricultural employers from discharging or otherwise discriminating against farmworkers for filing charges or testifying before the ALRB. The section has been interpreted broadly to protect employee attendance at an ALRB hearing.[49]

To establish a violation of section 1153(d), the ALRB General Counsel must prove four elements: a discriminatory act by the employer; employee participation in Board processes; employer knowledge of the employee's participation; and that the discrimination was motivated by the employee's participation in Board processes.[50]

Reducing an employee's hours of work after he testified against the employer in an ALRB backpay compliance hearing was found to violate section 1154(d) in *Abatti Farms, Inc.*[51] In *Bacchus Farms*[52] the discharge of three workers after attending an ALRB hearing in which two of them testified was found to be an unfair labor practice.

5. Section 1153(e): Employer Refusal to Bargain in Good Faith

ALRA Section 1153(e) prohibits employers and their successors from refusing to bargain in good faith with the elected union representative of their employees. Collective bargaining, the process of negotiating an agreement between management and the union representing workers, is the heart of the United States labor relations system. It is the mechanism through which workers and their representatives participate in establishing and administering rules of the work place, conditions of work, and compensation. The ALRA does not compel the parties to come to an agreement or to make specific concessions; it establishes the rules of the game, but then only requires the employer to bargain with the union representative in a *good faith effort* to come to an agreement. The ALRA and ALRB are not concerned with the substance of the agreement.

The difficulty in enforcing section 1153(e) is that "good faith bargaining" is hard to define. ALRA Section 1155.2(a) describes good faith bargaining as the mutual obligation of the employer and union to meet at reasonable times and "confer in good faith with respect to wages, hours, and other terms and conditions of employment" and sign a written contract if agreement is reached. Beyond this general statement, good faith is best described by the absence of bad faith bargaining. Proving violations, therefore, involves establishing the state of mind of the alleged violator.

a. Indicators of Bad Faith Bargaining

Although bad faith bargaining may occasionally be established by clearcut statements, such as "I'll never sign a contract with a union," more often it must be established through inference from external conduct. Indicators of bad faith bargaining range from blatant to very subtle types of conduct.

(1) Dilatory Tactics. There are a few obvious procedural indications of bad faith, such as when an employer refuses to meet with the union[53] or refuses to meet at reasonable times and places. Other *dilatory tactics* include delays in scheduling meetings, cancelling meetings or not showing up, or attaching preconditions to entering into negotiations. Similarly, an employer is not taking the bargaining obligation seriously if he sends a negotiator who has no authority to bind the party or switches negotiators several times, causing the negotiations to remain at a standstill while previous discussions are repeated for the newcomer. Refusal or delay in submitting proposals and counterproposals are other forms of dilatory tactics, as is withdrawing or changing proposals after agreement has been reached on them.

(2) Per Se Violations of the Bargaining Obligation. Certain types of conduct so clearly indicate bad faith bargaining that they have become known as *"per se" violations* of section 1153(e), and the state of mind of the party need not be proven. These acts include refusing to sign a written contract that has been agreed to by both parties; making unilateral changes in wages or working conditions; offering to negotiate directly with employees; and withholding information necessary for bargaining. An employer's unilateral changes in wages or working conditions after a bargaining representative has been selected effectively tells employees that even without collective bargaining they can secure advantages as great as, or greater than, those the union can secure. If an employer bypasses the union and attempts to bargain directly with employees, it is undermining the collective bargaining system by refusing to deal with the statutorily selected exclusive representative of the employees.

A union cannot bargain effectively for wages and benefits unless it has accurate and complete information on current wages and benefits. An employer's refusal or delay in providing requested information concerning individual earnings and wage rates, job classifications, merit increases, or pension plans makes effective collective bargaining impossible.

(3) Bargaining Table Misconduct. Even if the employer meets with the union and supplies requested information in a timely manner, certain *bargaining table behavior* can indicate bad faith bargaining. Refusal to discuss a mandatory subject of bargaining, including wages, hours, fringe benefits, or any other condition of employment, violates section 1153(e)[54]. Submitting obviously unacceptable proposals can indicate bad faith as can refusal to agree to trivial or noncontroversial items, such as union use of a company bulletin board.

Surface bargaining is the most subtle form of bad faith bargaining. It occurs when a party goes through the motions of good faith bargaining, but keeps the negotiations from reaching an agreement. Often difficult to detect, surface bargaining has led courts to remark that bad faith bargaining is still prohibited, even when done with sophistication and finesse.

One form of surface bargaining is simply to talk the other side to death.

> [T]o sit at a bargaining table, or to sit almost forever, or to make concessions here and there, could be the very means by which to conceal a purposeful strategy to make bargaining futile or fail...it takes more than mere "surface bargaining" or "shadow boxing" to a draw or "giving the union a runaround while purporting to be meeting for purposes of collective bargaining."[55]

Typically surface bargaining consists of a combination of actions which, when viewed as a whole, indicate that a party is merely going through the motions of bargaining. The bad faith conduct might include any of the actions listed above, or other types of insincere bargaining such as systematically opposing virtually all proposals submitted by the other side without providing an economic reason for such opposition, assuming inflexible positions on proposals, or making a take-it-or-leave-it proposal and breaking off negotiations. Unfair labor practices away from the bargaining table, such as discharging union supporters, may also be an indication of surface bargaining.

A respondent's defense to a surface bargaining ULP charge is usually that it was engaged in hard bargaining. The ALRB must then weigh all factors to determine if, when viewed as a whole, the employer's activities add up to bad faith bargaining or are sincere hard bargaining.[56] For example, in *Montebello Rose Co. and Mount Arbor Nurseries*[57] the ALRB noted that the two companies, which were engaged in joint bargaining with the UFW, met several times with the union, exchanged and discussed proposals, and reached agreement in some areas. Nevertheless, a full review of the facts led the ALRB to conclude that the companies' negotiations were a "charade or sham." Mount Arbor delayed in submitting a complete counterproposal, instituted two unilateral changes in wages, made a take-it-or-leave-it offer and abruptly declared that negotiations were at an impasse. At the bargaining table, both companies consistently opposed a dues check-off provision on the basis of cost, but they admitted they had made no estimate of the cost of deducting union dues from worker paychecks. The employers claimed to be representing their employees' best interests by opposing the good standing provision, a claim which rejects the union's statutory role as exclusive representative of employees. Montebello also attempted to bargain directly with employees. The ALRB concluded that both companies had engaged in bad faith bargaining.

b. Subjects of Bargaining

The ALRA requires the parties to bargain over "wages, hours, and other terms and conditions of employment,"[58] but the Act does not define what subjects fall under these headings.

The courts have defined three categories of bargaining subjects—mandatory subjects, permissive subjects and prohibited subjects. *Mandatory subjects* are those over which refusing to bargain is an unfair labor practice.[59] An exhaustive list of mandatory subjects of bargaining has been developed in NLRB, ALRB, and court decisions. Wage-related issues are mandatory topics and include hourly rates of pay, piece-rates, overtime pay, shift differentials, paid holidays, vacations, severance pay, regular bonuses, pension plans, profit-sharing plans, merit and wage increases, and company-provided meals and housing. Mandatory topics relating to hours include work schedules, premium pay and Sunday work. Other terms and conditions of employment which are mandatory subjects of bargaining include grievance and arbitration procedures, layoffs, discipline and discharge, workloads, sick leave, work rules, seniority promotions and transfers, retirement, union shop provisions (a requirement that workers must join the union after five days to keep their jobs) or agency shop provisions (a requirement that workers must pay the equivalent of fees and dues to the union), dues checkoff, hiring hall rules, plant rules, health and safety procedures, management rights clauses, no-strike clauses, subcontracting and sometimes mechanization.

Permissive subjects of bargaining are those that either party may refuse to bargain over without committing an unfair labor practice. Some examples are plant locations, corporate organization, or the size of the supervisory force.

Agreements concerning *prohibited subjects* of bargaining will not be enforced by the ALRB or the courts. Examples included a closed shop provision (a worker must be a union member before being hired) or provisions that violate a law.

c. Bargaining Over Business Decisions

A current ALRB controversy is whether decisions that affect the business operation and have a substantial impact on the work force are mandatory subjects of bargaining. For example, must a farm which signed a union contract while it was growing labor-intensive lettuce bargain with the union over a decision to switch to mechanically-harvested sugar beets? Ultimately, the question of switching crops requires weighing the value of the collective bargaining process (and the union's interest in protecting members' jobs) against the burden collective bargaining places on the employer's freedom to conduct his or her business. The NLRB, the ALRB, and the courts have developed a procedure to determine which business decisions must be negotiated with the union.

The procedure begins with two clear-cut principles which are at opposite ends of a continuum (Figure 6.1). On one end is the rule that an employer decision to subcontract or contract out bargaining unit work and thus reduce union employment is a mandatory subject of bargaining. In its 1964 opinion in *Fibreboard Corp. v. NLRB*,[60] the U.S. Supreme Court noted that the company's decision to subcontract its maintenance work resulted in the termination of work for the whole bargaining unit, so the decision clearly affected the "terms and conditions of employment." Since the business decision replaced union workers with nonunion workers and did not alter the company's basic operation or involve new capital investment, requiring the employer to bargain about the issue would not significantly abridge his freedom to manage his business.

At the other end of the continuum is the Supreme Court's ruling that an economic decision to partially close a business is generally not a mandatory subject

Figure 6.1

Duty to Bargain Over Business Decisions

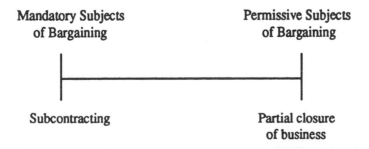

Mandatory Subjects
of Bargaining

Permissive Subjects
of Bargaining

Subcontracting

Partial closure
of business

of bargaining. The Supreme Court reasoned that the burden placed on the conduct of business by having to bargain over whether part of the business should be closed outweighs the benefits of encouraging collective bargaining, even if the decision has a substantial impact on employment.[61]

Two additional principles enter into this reasoning. First, an employer action which results in the termination of employment for one or more workers, such as the partial closing of a business or a runaway shop (closing a union business and opening a nonunion business nearby) which is motivated by anti-union animus, violates NLRA section 8(a)(3) and ALRA section 1153(c).[62] The remedy for anti-union business changes is an order to reinstate the discharged workers in other parts of the business. Second, the decision to close an entire business, even if the employer announces that he or she is going out of business "to get rid of the union," ends the employer-employee relationship and is not a mandatory subject of bargaining.[63] Thus, changes in an ongoing union business come under scrutiny, but business closures do not.

A number of employer actions fall somewhere along the continuum between subcontracting, which requires bargaining, and closure, which does not. When such business decisions are challenged by unions and workers, analysis begins with the question of whether the employer action eliminates bargaining unit jobs. If no jobs are eliminated, then there is no duty to bargain over the decision. If jobs are lost, then the analysis asks whether the decision changed the scope and direction of the business. If the decision does not change the basic organization of the business, such as a decision to switch from picking into bags to placing melons on a conveyor belt in order to get them into a truck, then the employer must submit the issue to collective bargaining because the benefits of collective bargaining outweigh the burdens on business of negotiating the change.

Decisions which involve a change in the basic scope and direction of the business, such as switching from growing labor-intensive vegetables to growing cotton, are considered to be related closely to the decision about whether to be in business at all and are left to management prerogative. While there is no duty to bargain over such *decisions* (decision bargaining), employers must bargain over the *effects* of decisions which alter the scope of the business (such as changing crops) if such decisions affect wages, hours, or working conditions. Similarly, employers must bargain over the effects of business decisions that do not eliminate jobs but do alter wages, hours, or working conditions.

The resolution of bad faith bargaining charges against employers who make decisions which affect their businesses without bargaining is illustrated by the decision tree in Figure 6.2. First, economically-motivated decisions must be separated from discriminatory conduct, so an initial "no" answer to the anti-union question moves the case to the second level. At this level, the question is whether the business decision resulted in a loss of jobs. In *Mount Arbor Nurseries, Inc. and Mid-Western Nurseries, Inc.*,[64] the ALRB found that the employer's decision to

Figure 6.2

Is the Business Decision a Mandatory Subject of Bargaining?

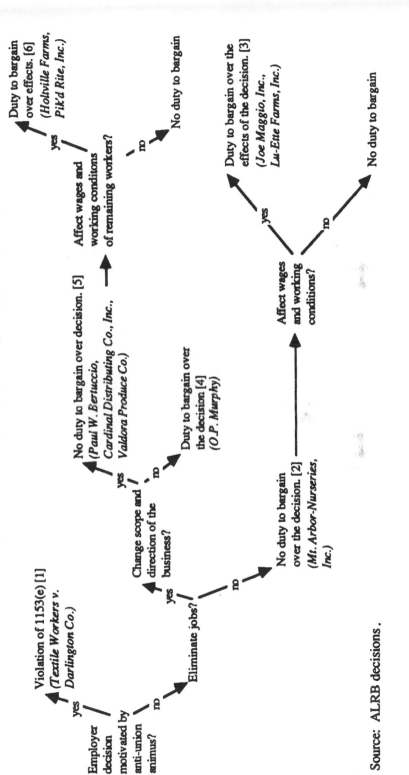

Source: ALRB decisions.

grow almonds instead of roses had no impact on the number of jobs in the bargaining unit, so the employer was not required to bargain over the decision. (Result 2 on Decision Tree.) However, the decision to contract out the work in almonds was a mandatory subject of bargaining and the employer violated section 1153(e) by not bargaining with the certified union over the subcontracting.

In *Joe Maggio, Inc.*[65] the employer adopted a new lettuce wrapping technology without giving the union an opportunity to bargain over the change. The workers had been cutting lettuce by hand and packing it in boxes in the fields; after the change, workers continued to cut the lettuce by hand, but they then put the lettuce on a conveyor belt which carried it to a machine where it was wrapped and packed by another group of employees. The ALRB reasoned that the new operation did not eliminate jobs and thus the employer did not have to bargain over the decision to switch to the lettuce wrapping technology. However, the decision did change wages and working conditions, so the employer had a duty to bargain over the effects of the decision.[66] (Result 3).

If the employer decision eliminates jobs, another question is raised: does the decision change the scope and direction of the business? In *O. P. Murphy Produce Co., Inc.*[67] the ALRB found that the employer decision to mechanize the fresh market tomato harvest was a mandatory subject of bargaining because it would result in the elimination of about 80 percent of the bargaining unit jobs, and the business was substantially unaltered in scope or direction. (Result 4). The company was still growing fresh market tomatoes, so the decision was not like starting or ending a product line or going out of business. Here, the burden on the employer imposed by collective bargaining was outweighed by the benefits of collective bargaining as the preferred method of resolving workplace disputes, so the employer had to bargain over the mechanization decision.

The opposite conclusion was reached in *Paul W. Bertuccio*[68] where the employer decided to sell his garlic crop for seed rather than harvest it as usual. The ALRB likened Bertuccio's decision to a termination of part of his business. Despite the fact that the decision resulted in a loss of jobs, the ALRB reasoned that the scope of the business changed enough so that there was no duty to bargain. (Result 5). The same reasoning was used in *Cardinal Distributing Company, Inc.*[69] to determine that an employer's decision to grow or discontinue a certain crop is not a mandatory bargaining subject. Similarly, in *Valdora Produce Company*[70] the ALRB reasoned that decisions over which crops to grow involve the scope and direction of the business, and employers can make such decisions without bargaining

An employer decision which affects the remaining workers' wages and working conditions generates an employer obligation to bargain over the effects of the decision. (Result 6). In *Holtville Farms, Inc.*[71] the Board concluded that the employer violated section 1153(e) by failing to bargain over the effects of its decision to stop growing lettuce. Similarly, a decision to cease operations is management's sole perogative, but the employer has a duty to bargain over the effects of that decision.[72] However, unions typically have little bargaining power over the effects of such business decisions; a strike called to protest the switch from lettuce to alfalfa is likely to have little effect because only a few equipment operators and irrigators are employed in alfalfa.

d. Impasse

After a union is certified as the bargaining agent for a group of workers, the employer is required to bargain until a contract has been reached or there is an impasse. A bona fide impasse exists when the parties are unable to reach an agreement despite good faith efforts to do so. The exact time that impasse occurs is critical; after a bona fide impasse exists, an employer can lawfully implement its final wage and benefit offer unilaterally. If bargaining has dragged on for several years, the employer may be anxious to declare an impasse and raise wages in order to attract a work force.

Negotiations are not at an impasse if the parties still have room to bargain in an unresolved area even if they are deadlocked in some areas. In such a case, continued negotiations on topics where there is still room for bargaining can loosen or dislodge the deadlock in other areas. For example, in *Montebello Rose Co., Inc.* and *Mount Arbor Nurseries*[73] the parties appeared to be deadlocked over the issues of union security and dues checkoff, but there had been no in-depth discussion of wages. The employer declared an impasse and presented its "final offer." The ALRB held that there was no legitimate impasse since little discussion of wages had taken place and the parties had been offering counterproposals on the topic. The premature declaration of impasse aborted the negotiating process while there was still room for movement and therefore was bad faith bargaining.[74]

A bona fide impasse does not exist if the deadlock is caused by one party's bad faith bargaining. While the ALRB cannot impose contract terms on the parties, it can examine the reasonableness of the parties' positions.[75] Looking at the impasse between the parties in *Montebello Rose*, the ALRB concluded that the "deadlock" resulted from the employer's bad faith bargaining. The employer had advanced a patently improbable justification for refusing to agree to a good-standing provision in the contract; it claimed to be protecting employees from the UFW, the union that employees had chosen in an election. The employer was bargaining in bad faith by rejecting a basic principle of collective bargaining: the certified union is the *exclusive* representative of the employees. The employer also rejected a dues check-off system because of the "high costs" of changing the payroll system, but made no serious effort to estimate the cost. After examining each of these employer arguments, the ALRB concluded that the "impasse" was the result of the employer's bad faith bargaining.

e. Continuing Duty to Bargain

Once a union is certified as the exclusive representative of employees, the employer has a duty to bargain in good faith until a contract (or impasse) is reached. Thus, the employer cannot make unilateral changes in wages or working conditions while the contract is in effect. If the employer wants to make such a change, it must give the certified union the opportunity to bargain over it.

An employer's refusal to bargain with a union because of its belief that the union had lost majority support is a violation of 1153(e). This is true even if the employer has taken an employee poll;[76] or if the employer believes in good faith that the union has lost majority support.[77] If the employer claims that the certified union has lost majority support from the workers and refuses to bargain, the employer does so at his own risk. Only a certified decertification election or the certified election of another union representative can prove that a union has lost majority support among employees. The union can charge the employer with violating section 1153(e) for failing to bargain with a certified union.

B. UNION UNFAIR LABOR PRACTICES

When originally passed, the National Labor Relations Act, or Wagner Act, enumerated only unfair labor practices that employers could commit. Public concern after World War II about the power of "big labor" led to the Taft-Hartley Act of 1947, which amended the NLRA to include union unfair labor practices.

The ALRA contains an enumeration of union unfair labor practices similar to the NLRA. The major differences are the right of a farmworker union to require members to be in good standing in order to remain employed on a union farm and a qualified rather than absolute limitation on secondary boycotts.

1. Section 1154(a): Restraint or Coercion

ALRA Section 1154(a)(1) prohibits unions from restraining or coercing employees in the exercise of their section 1152 rights to join a union, engage in concerted activity, bargain collectively, or refrain from these activities. Union responsibility for misconduct must be proven by workers who allege that their rights were violated. Therefore, the evidence must show either that the person who engaged in unlawful conduct was an agent of the union or that repeated coercive acts occurred in a situation over which the union had control, such as a picket line.

Violence or threats of violence constitute coercion and restraint within the meaning of section 1154(a)(1). Mass union picketing that blocks ingress and egress to an employer's property is also unlawful, since it tends to coerce or intimidate employees in the exercise of their rights. It is also coercive and therefore unlawful for union agents to follow nonstrikers away from the worksite or to give the impression that they are writing down nonstrikers' license numbers. Likewise, threatening employees with loss of employment for engaging in protected activities in unlawful.

Section 1154(a)(1) is also violated by union misconduct directed toward nonemployees if it is within the view of employees or is likely to come to their attention. In *Western Conference of Teamsters (V. B. Zaninovich and Sons, Inc),*[78] Teamsters business agents violated 1154(a)(1) by damaging the employer's property and threatening supervisors with bodily harm in the presence of employees, acts which might encourage employees to support the Teamsters in order to avoid similar violence against them. Other unlawful acts in this case included union agents throwing rocks, bottles, and other objects at workers and supervisors in the presence of other workers, breaking car windows, throwing Tijuana (3-cornered) tacks in front of cars, forcing workers' cars off the road, following workers in a menacing manner, and threatening them with bodily harm.

In *United Farm Workers of America* (Marcel Jojola)[79] the ALRB dealt with the issue of union coercion through residential picketing. Approximately 50 union pickets marched along the sidewalk outside the homes of two workers, chanting slogans and shouting epithets at the residents for working during a strike. Picketing began before dawn at a third nonstriker's home. The ALRB concluded that such union actions had a tendency to coerce or restrain workers by denying them the right to work during a strike. However, the Board did not prohibit all residential picketing; in a subsequent case, residential picketing of strikers homes was allowed under certain restrictions.

Section 1154(a)(2) prohibits a union from restraining or coercing an employer in the selection of representatives for collective bargaining or the adjustment of grievances. This section prohibits a union from saying that it will not bargain with a particular negotiator.

2. Section 1154(b): Union Discrimination Against Employers or Union Inducement of EmployerDiscrimination

Section 1154(b) of the ALRA prohibits a union from discriminating against a union member or causing or attempting to cause an employer to discriminate against a union member in violation of 1153(c) if the worker has been unjustly denied membership in the union. In other words, a union is prohibited from requiring an employer to discharge an employee who, for example, is challenging union leaders.

Most collective bargaining agreements include a union security clause through which the union attempts to maintain its strength. ALRA section 1153(c) authorizes the inclusion of a union shop provision in a collective bargaining agreement which requires all new workers to join the union within five days and maintain their union membership "in good standing" to keep their jobs.[80] Under the NLRA, good standing means that all union members must pay union dues, fees, and assessments in order to remain union members. Under the ALRA, a union shop agreement can require union members to abide by whatever reasonable rules the union adopts. The union can enforce the good standing rule by requiring an employer to discharge an employee who is no longer a union member in good standing.[81]

The UFW has more than 30 rules which must be followed to be a UFW member in good standing, and the UFW is regularly accused of using these good standing rules to keep members "in line." However, the UFW must provide due process to all members who are accused of being out of good standing; it cannot terminate UFW membership except in compliance with its constitution and bylaws that give all members full and fair rights to speech, assembly, and equal voting and membership privileges when UFW rules are adopted.[82]

The ALRA good standing rule has been controversial because the UFW has a mandatory Citizen Participation Day fund (CPD). CPD is a UFW contract provision under which workers are paid for one day when they do not work and the funds are remitted directly to the Citizen Participation Committee of the UFW. CPD funds are allocated to two programs, one for social welfare programs for UFW members and their families, and the other to a political action committee.

In *United Farm Workers of America*,[83] Jesus Conchola asserted that he did not want his funds to be used for political purposes, but feared that if he objected, the union would find him not in good standing and have him discharged from his job. U.S. Supreme Court rulings prohibit unions from financing expenditures which are not germane to collective bargaining with mandatory contributions from employees who object to such expenditures.[84] Unions may require contributions from all members to help defray the expenses of contract negotiations and administration, including the expenses of resolving grievances, lobbying efforts, or other political activities closely related to collective bargaining.   However, mandatory contributions cannot be used for purely political purposes such as campaigning for a particular political candidate. In the Conchola case, the UFW was ordered to allow employees who object to the use of their funds for political or ideological purposes that are not related to collective bargaining to receive a rebate of that part of their contributions which would be used for these purposes.

In a subsequent arbitration case at the Mann Packing Company, the arbitrator held that the rebate program must allow the employee to deliver the rebate request by hand to the union and that the rebate must remain in effect until revoked by the employee. The U.S. Supreme Court in a 1984 case held that a union cannot collect full dues and then refund the disallowed part later if there are acceptable alternatives, such as a reduction of dues for objecting members or the placement of mandatory contributions in an interest-bearing escrow account.[85] The ALRB has not yet indicated whether it will apply this Supreme Court ruling to agricultural unions.

The broad ALRA good standing clause gives a farmworker union a great deal of influence over a worker's employment on a union ranch. Thus, an important aspect of a good standing clause is that union members must be given due process in union disciplinary hearings which could result in loss of good standing. Due process includes the right to receive written charges; advance notice of a hearing with a reasonable time to prepare a defense; and a full and fair hearing.[86] In *Pasillas v. ALRB*,[87] the court of appeal held that the union had the power to suspend members for strikebreaking, but that the union had violated section 1154(b) by denying due process in the disciplinary hearings.

3. Section 1154(c): Bad Faith Bargaining

ALRA section 1154(c) requires a union to bargain in good faith with the employer after it has been certified as the elected representative of the employees. The same standards for employer good faith bargaining apply to unions. In 1986, for example, the UFW was found to have bargained in bad faith by cancelling meetings, delaying negotiations on five separate occasions, and by failing to submit counterproposals to employer Maggio.[88] The ALRB reasoned that the unavailability of the UFW's representatives because they were busy with other union businesses is not a defense to a bad faith bargaining charge.

4. Other Union Unfair Labor Practices

Sections 1154(d), (g), and (h) prohibit unions from engaging in certain types of strikes, boycotts, and pickets. Unions are prohibited from charging excessive or discriminatory fees to potential members by section 1154(e). In deciding what is excessive, the ALRB must consider the practices and customs of other agricultural unions and current agricultural wages. Finally, section 1154(f) prohibits unions from "featherbedding," that is, requiring employers to pay workers for services not performed.

C. REMEDIES

ALRA section 1160 sets out the authority of the ALRB to issue remedies for unfair labor practices. It empowers the ALRB:

to issue an order requiring [the violator] to cease and desist from such unfair labor practice, to take affirmative action including reinstatement of employees with or without backpay, and making employees whole, when the board deems such relief appropriate for the loss of pay resulting from the employer's refusal to bargain, and to provide such other relief as will effectuate the policies of this part.

The ALRB's remedial authority is broader than that of the NLRB because the ALRA specifically authorizes the makewhole remedy for bad faith bargaining as well as "such other relief as will effectuate" the purposes of the ALRA. ALRB and NLRB remedies cannot be punitive (such as levying fines for violating the ALRA) and can only be used to restore the status quo before the violation.

The standard set of remedies applied by the ALRB to most unfair labor practice findings includes a cease and desist order (an order telling the violator to stop engaging in specific misconduct), and the posting, mailing, and reading of a Notice to Workers informing them of the outcome of ALRB proceedings. Reinstatement and backpay are the usual remedies for discriminatory discharge. Makewhole is an additional remedy unique to the ALRA for bad faith bargaining. Expanded access is sometimes ordered to remedy an employer's denial of access to a labor union.

1. Notice to Workers

The usual ALRB remedy includes a Notice to Workers explaining in nontechnical language the rights given to them by the ALRA and the fact that a hearing has taken place before the ALRB, that the employer or union was found to have violated the ALRA, and that the employer or union agrees to stop violating the ALRA. The Notice lists the specific conduct which the employer or union will no longer engage in and it is signed by the employer or union. The Notice must be posted at the employer's premises for a specified time period; sent to all workers who were employed when the violation occurred; and distributed and read to current employees in English and Spanish or other appropriate languages during work time by a company or union representative or an ALRB agent. After the reading of the Notice, employees are given an opportunity to ask questions about the Notice and their rights under the ALRA.

The ALRB found that the reading requirement is necessary because a substantial part of the farm work force is illiterate.[89] The mailing requirement is deemed necessary because many workers who were present when the violation occurred may not be working for the same employer when the notice is read, and it is important that they too know that the violation has been remedied. The ALRB has determined that distribution of the Notice is required so that workers can review their rights and the outcome of the ALRB proceedings in the privacy of their homes.

2. Expanded Access

When the Board finds that an employer has violated the access rule, it seeks to restore to the workers the lost opportunity to talk to organizers. The Board has also ordered expanded access to counteract the effects of the employer's failure to provide an adequate prepetition list.[90] Types of expanded access include an additional organizer for every 15 employees in a crew during the regular access period and, in especially egregious cases, allowing the union an hour to talk to workers during regular work time (i.e., at the employer's expense).[91]

3. Reinstatement

The Board ordinarily orders reinstatement and backpay for a worker who has been discriminatorily discharged, demoted, or transferred in violation of section 1153(c) or 1154(b). The aggrieved worker is entitled to an unconditional and unambiguous offer of reinstatement to the former job, even if he or she has obtained substantially equivalent employment elsewhere[92] and even if the employer is required to terminate a replacement worker in order to reinstate the aggrieved employee. Since the objective is to restore the discriminatee to the status quo, the employer must offer the discriminatee a substantially equivalent job even if the discriminatee's original job no longer exists. To be substantially equivalent, the new job must have similar wages, hours, duration of employment, fringe benefits, duties and responsibilities and working conditions and require similar job skills. If the new job involves lower pay, less desirable work, a different job classification with different duties, a different location, or a seasonal job (if the original job was year round), the ALRB has found that the job was not substantially equivalent and, therefore, the ALRB's remedial order had not been complied with.[93]

4. Backpay

The party which caused an employee to be discriminatorily discharged is normally required to restore the worker to the same financial position he or she would have been in if he or she had not been discharged. Accordingly, the general formula for determining the daily amount of backpay (B_p) is the amount the worker would have earned had he or she not been discharged (W) less the interim earnings (W$_i$) plus job seeking expenses (E) and interest (I):

$$B_p = W - W_i + E + I$$

The interest rate is that set by the Internal Revenue Service for delinquent taxes.[94] After the ALRB decides that a discriminatory discharge has occurred, a separate hearing is held to determine the amount of backpay. At this hearing, an ALRB agent recommends an amount of backpay and the violating party (employer or union) can present evidence disputing it.

The actual computation of backpay can be exceedingly complex because of seasonality and migrancy. The ALRB uses two methods to compute backpay: the average daily wage method and the representative employee method. The average daily wage method was used in *Maggio-Tostado, Inc.*[95] where the eight discriminatees had worked at a piece-rate wage for the respondent. Maggio's work force was characterized by large and fluctuating numbers of employees, high turnover, and no discernible seniority system for layoff and rehiring. The ALRB computed the average daily wage of the discriminatees by dividing the total number of hours worked each day by the number of employees who were working and then multiplying these average daily hours by the appropriate hourly wage to obtain an average daily wage. Daily gross backpay was calculated for the number of days in which more than eight employees were working, since it was on these days that the workers who were discriminated against would presumably have worked.

The representative employee method was used in *Butte View Farms*. In that case, the discriminatee was a tractor driver/irrigator. His backpay was calculated by using the wages received by another employee whose work was similar to that of the discriminatee.

Because of the day-to-day fluctuations in agricultural employment, the backpay formula ($B_p = W - W_i + E + I$) is calculated on a daily basis when sufficient information is available. For every day the discriminatees would have worked, gross expected wages are calculated (say $100). Actual earnings from another job are subtracted (say $60), yielding a daily backpay award ($40). Then job hunting expenses and interest are added to this daily amount. No deduction is made for days when a worker's earnings exceed what the worker would have earned ($100) or for earnings on days the workers would not have been employed. The ALRB has rejected a formula based on a longer period of time, such as the NLRB quarterly computation formula, as not appropriate in agriculture because of the seasonal nature of employment.[96] Under a longer computation period, a worker's peak harvest earnings in another job might reduce or eliminate any backpay owed, so the discriminatee would be deprived of the earnings that would have been earned on the violator's farm while the worker was unemployed.

5. Makewhole

Section 1160.3 of the ALRA authorizes the ALRB to issue a makewhole order which requires an employer to compensate workers for an employer's refusal to bargain in good faith. Experience under the NLRB convinced many experts that this remedy is necessary to stop an employer from continuing to bargain in bad faith and to shift the monetary losses of bad faith bargaining from the employees to the employer who engaged in the illegal conduct. A former chairman of the NLRB summed up the need for a more adequate remedy this way:

> Every [NLRB] member... conceded the inadequacy of the Board's 8(a)(5) remedies. The losses to employees, especially in first bargaining situations, who are deprived for 1, 2, or sometimes many more years of their right to be represented are palpable. The weakening of their bargaining agent's status is admitted. The savings to respondent employers from delaying the onset of bargaining for these long periods can be enormous. Until this basic profit from unfair practices is removed, the incentive to mock the statute's promises with lengthy delays is apparently compelling.[97]

The NLRB does not issue makewhole orders, however, because without clear authorizing language in the NLRA, the NLRB's authority to do so is in question.[98]

The Labor Law Reform Act of 1977 would have authorized a makewhole remedy for bad faith bargaining under the NLRA. Workers would have been awarded an amount equal to the average wage negotiated at similar plants where collective bargaining proceeded lawfully, using data from the U.S. Bureau of Labor Statistics to determine the amount owed. The 1977 Labor Law Reform Act passed the U.S. House of Representatives, but was blocked in the Senate.

Without NLRB precedent for guidance, the ALRB has had to determine (1) when to apply the makewhole remedy, (2) how to calculate its amount and the duration, and (3) who should receive makewhole monies. The ALRA directs the ALRB to order a makewhole remedy when it "deems such relief appropriate."[99] The Board issued its first makewhole orders in *Adam Dairy*[100] and *Perry Farms*.[101]

In *Perry Farms* the Board decided that makewhole was appropriate any time workers had lost wages as a result of any employer's unlawful refusal to bargain.

> The employees suffer this same loss whether or not the employer's refusal to bargain is designed solely to procure review in the courts of the underlying election issues or is of the flagrant or willful variety.[102]

This means that in a technical refusal to bargain (when the employer refused to bargain in order to obtain judicial review of the Board's certification of the union), if the employer lost the appeal, the employer would be liable for makewhole. (If the employer won it would not have a duty to bargain with the union.) In 1979, in *J. R. Norton*,[103] the California Supreme Court ruled that the ALRA did not allow such a blanket makewhole remedy, and ruled that makewhole was not appropriate where the employer had challenged the certification in good faith. Since then the ALRB has developed a two-pronged test to determine when makewhole is appropriate in technical refusal to bargain cases. First, the Board looks to see if the litigation posture of the employer is reasonable. If it is, the Board then determines whether the employer was challenging the election in good faith or if the employer was simply trying to delay the obligation to bargain. If the challenge was in good faith, makewhole is not appropriate.[104]

In a 1987 case, a state court of appeal ruled that makewhole could not be imposed unless the Board first determined that a contract with higher pay would have been negotiated but for the employer's bad faith or refusal to bargain.[105] The court ordered the Board to use the same test as is used in unlawful discharge cases— the ALRB General Counsel must first prove that the employer bargained in bad faith or refused to bargain, then the burden of persuasion shifts to the employer, who must show that even if bargaining had proceeded in good faith, no contract with higher pay would have been signed.[106]

The Board has also ruled that not all unilateral changes in wages and working conditions, although violations of 1153(e), warrant makewhole awards. The Board has ruled that makewhole is not an appropriate remedy for a union's failure to bargain in good faith.[107]

The ALRB has attempted to follow the calculation formula proposed in the 1977 Labor Law Reform Act. The 1977 Labor Law Reform Act provided that makewhole should be calculated as the difference between (i) the wages and other benefits actually received by employees during the period of employer bad-faith bargaining and (ii) the wages and fringe benefits other workers employed at companies which had bargained lawfully were receiving. This difference is multiplied by the percentage change in wages and other benefits reported by the Bureau of Labor Statistics (BLS). This comparison with other contracts avoids additional litigation on the issue of whether or not the employer would have reached a contract or agreed to a particular provision.

In *Adam Dairy* the Board first announced a formula to calculate makewhole remedies. Since BLS does not publish statistics on union wages and fringe benefits in agriculture, the ALRB decided that it would base the makewhole wage rate on the average basic wage rate ($3.13) in the 37 UFW contracts negotiated pursuant to ALRB certifications. (UFW contracts were used since the UFW was the complaining union.)

Next, the ALRB had to determine a method to compute the value of fringe benefits such as health insurance, pensions, and paid vacation. Seeking a generalized approach to avoid litigation over specific provisions, the ALRB relied on fringe benefit data from the Bureau of Labor Statistics' *Employee Compensation in the Private Nonfarm Economy* (1974) to determine that wages constituted 78 percent of the total worker compensation received in the nonmanufacturing sector. Thus, gross makewhole compensation was calculated to be $4.01:

$$.78 \text{ (gross makewhole)} = \$3.13 \text{ hourly wage}$$

$$\text{Gross makewhole} = \frac{3.13}{.78} = \$4.01 \text{ per hour}$$

The employer was ordered to pay each worker the difference between $4.01 and the amount actually paid in wages and benefits during the makewhole period. Workers earning 10 percent more than the basic wage were entitled to a makewhole wage 10 percent larger, e.g., $4.41.

Since *Adam Dairy* was decided in 1978, there has been much litigation over the ALRB's method of computing makewhole awards. During the makewhole period in *Adam Dairy*, 30 of the 37 contracts had the same general labor wage. The contracts were relatively similar because they involved two commodities: vegetables and tree fruits. However, none of the contracts covered a dairy.[108]

In 1978 the Board in *J. R. Norton* changed its method of calculating makewhole from averaging all contracts to looking only at comparable contracts in order to determine the average makewhole wage (the fringe benefit factor remained at 22 percent).[109] The Board reasoned that the variance in wages from region to region and from commodity to commodity had increased since the makewhole period in *Adam Dairy*. The factors to be used to determine which contracts were comparable were the time frame of the contracts, the size of the work force, the commodity, and the location of the employer. The ALRB stated that all comparable contracts should be used, but in two cases, a single comparable contract was sufficient.[110] The comparable contract method generated litigation over exactly which contracts were comparable; litigation over comparable contracts was not necessary with the *Adam Dairy* method.

In 1983, the Board reduced the makewhole fringe percentage from 22 to 15.7 percent.[111] The 22 percent figure included mandatory employer contributions for social security and worker's compensation, which are not subject to collective bargaining, so the Board excluded them from the makewhole calculations. According to the 1974 BLS survey, such mandatory contributions were 6.3 percent of total compensation in nonmanufacturing, yielding a 15.7 percent modified fringe benefit percentage. In 1984 the Board reconsidered *J. R. Norton*, and devised a new way to calculate the value of fringe benefits in union contracts.[112]

UFW labor contracts usually include vacation pay, paid holidays, rest periods, overtime pay and night time premiums. The Board decided to calculate the dollar value of fringes available under the average UFW contract to each individual worker.[113] For example, workers usually must work a minimum number of hours to be eligible for vacation pay, and vacation pay is usually calculated as a percentage of annual earnings, so a worker with five years seniority who worked 800 hours during the year may be eligible for vacation pay that is 4 percent of gross earnings. The *J. R. Norton* method determines *if* the employee would be eligible for vacation, and then calculates *how much* he or she would receive. Such calculations are made for each worker, versus the *Adam Dairy* method of a single percentage applicable to all workers.

Individual worker calculations are very tedious. For example, rest periods of 10 minutes for every four hours are required by labor laws; if comparable union contracts provide an extra five minute rest period, this extra rest time adds 2.08 percent of hourly wages (5 min + 240 min. = 2.08%). Each holiday was estimated to be worth 0.32 percent of hourly wages (1 day of 312-day year). Hourly contributions to pension plans and health insurance are added to the total award.

The gross makewhole calculation procedure for each worker involves two steps: first, calculate the average wage in comparable contracts ($5) and multiply it by the number of hours worked by each person (1,000 hours or $5,000). Then, multiply the employer's average contributions for each hour worked ($0.75) times the number of hours worked (1,000 or $750). Then add 0.32 percent times the hourly wage ($5) times the number of paid holidays provided in comparable contracts (6) and hours worked (32% x 6 days x $5 x 1,000 hours = $96). Then, multiply the rest period percentage (2.08%) by the hourly wage ($5) times hours worked ($2.50 or 1.5 times the hourly wage) times the number of overtime hours included (10 hours or $25). Finally, multiply the vacation pay percentage determined by seniority (2 percent) by the worker's annual earnings (.02 x $5,000 = $100). Adding up these numbers yields the worker's gross makewhole award ($6,075). Fringes are 17 percent of gross makewhole pay.

After calculating each worker's makewhole pay, the Board must subtract the actual wages and fringes received during the makewhole period by each worker who is owed an award. The makewhole award minus actual wages is the net makewhole award paid to the worker.

These individual worker calculations are time-consuming and expensive, since the ALRB must carefully examine employer payroll records. The ALRB continues to search for an efficient and equitable way to calculate appropriate makewhole awards. For example, using reasonable assumptions about the value of fringe benefits under UFW contracts, it appears that fringe benefits actually add 14 to 17 percent to the hourly wages of unionized California farmworkers.

In *Adam Dairy* the ALRB held that the makewhole period should extend from the employer's first refusal to bargain until the respondent begins to bargain in good faith. However, ALRB surface bargaining litigation is still wrestling with the problem of trying to establish the exact dates for the beginning and ending of the makewhole period, since the dates when bad faith bargaining begins and ends are not easily determined.

When a union strikes to protest bad faith bargaining and replacement workers are hired, the ALRB must determine which group or groups of workers should receive makewhole pay. In *Bruce Church, Inc.*[114] the Board ordered the company to award makewhole to current employees and to the strikers who had not been permanently replaced. Strikers do not receive base pay or benefits for the period while they were on strike, but they do receive the difference between what they would have earned if they had been working and the makewhole wages and benefits. Temporary replacement workers do not receive a makewhole award.

D. SUMMARY

Section 1152 of the ALRA grants farmworkers the right to organize and join a union and bargain collectively, or to refrain from union activities. Employer or union conduct which violates the rights created by the ALRA are unfair labor practices. Section 1153 enumerates employer ULPs, and Section 1154 enumerates unions ULPs.

Section 1153(a) prohibits employers from interfering with the worker rights enumerated in Section 1152. It protects workers engaged in protected concerted activities not involving a union, and prohibits employers from coercing employees during an election. Section 1153(b) prohibits employers from assisting a union or from favoring one union over another in an election, and section 1153(c) prohibits employers from discriminating against union supporters. The ALRB follows NLRB precedent in using a four-part test to determine if the employer has unlawfully discriminated against a worker.

Section 1153(d) protects workers from employer retaliation for participation in ALRB hearings. Finally, section 1153(e) creates a duty on the part of the employer to bargain in good faith with a certified union until an agreement or impasse is reached.

Section 1154(a) prohibits unions from coercing workers who are exercising their section 1152 rights to join or refrain from union activity. In order to prove a section 1154(a) violation, the ALRB General Counsel must show that a union agent was responsible, or that a violation occurred in a situation under union control. Section 1154(b), combined with section 1153(c), regulates a union's ability to require an employer to discharge a worker for failure to maintain good standing in the union. Section 1154(b) regulates the reasons a union can find a member not in good standing, and the procedures the union must follow to eject a member. Other parts of section 1154 regulate union bad faith bargaining, strikes, boycotts, and pickets.

The ALRB has broad remedial powers to compensate workers for ULPs, including reinstatement with backpay, expanded access and the makewhole award. The ALRB does not have the power to impose fines or penalties on employers; instead, it attempts to restore the status quo to workers.

NOTES

1. Cal. Lab. Code §1140.2 (West Supp. 1987).
2. California Senate Bill No.1, June 5, 1975.
3. (1978) 4 ALRB No. 25 rev'd on other grounds, 86 Cal. App. 3d 448 (3d dist. 1978).
4. (1977) 3 ALRB No. 51.
5. Arnaudo Bros., Inc. (1977) 3 ALRB No. 78.
6. (1977) 3 ALRB No. 67.
7. E. and J. Gallo Winery (1981) 7 ALRB No. 10.
8. Merzoian Brothers Farm Management (1977) 3 ALRB No. 62.
9. (1977) 3 ALRB No. 71.
10. Tom Bengard Ranch, Inc. (1978) 4 ALRB No. 33.
11. Valley Farms (1976) 2 ALRB No. 41.
12. Coachella Imperial Distribution (1979) 5 ALRB No. 73 (union organizers have the right to enter a field one hour after work ends, even when employees leave work in shifts).
13. Jack Pandol and Sons (1977) 3 ALRB No. 29, rev'd and remanded on other grounds, 98 Cal. App. 580 (5th dist. 1979).
14. Cal. Admin. Code tit. 8, §20910 (1977).
15. Henry Moreno (1977) 3 ALRB No. 40; Tenneco West, Inc. (1977) 3 ALRB No. 92.
16. Foster Poultry Farms (1980) 6 ALRB No. 15; Miranda Mushroom Farms, Inc. (1980) 6 ALRB No. 22.
17. 370 U.S. 9 (1962).
18. NLRA section 10(e) is equivalent to Cal. Lab. Code §1160.3.
19. Pappas & Company (1979) 5 ALRB No. 52.
20. See NLRB v. Washington Aluminum Co., 370 U.S. at 17 (1962).
21. U.S. Congress, Senate, Committee on Education and Labor, Report on the Wagner Act, 74th Congress, 1st Sess., 193, pp. 9-11.
22. Louis Caric & Sons (1978) 4 ALRB No. 108.
23. See Miranda Mushroom Farms, Inc. (1980) 6 ALRB No. 22, where the Board found that the employer gave unlawful assistance by encouraging the formation of an employee organization and permitting it to use company facilities

while denying the same treatment to the rival union, but the Board stopped short of finding that the employer dominated the organization.

24. (1978) 4 ALRB No. 82.

25. (1979) 5 ALRB No. 31, aff'd, 101 Cal. App. 3d 826 (4th dist. 1980).

26. (1981) 7 ALRB No. 10.

27. Louis Caric & Sons (1978) 4 ALRB No. 108.

28. NLRB v. Cabot Carbon Co., 360 U.S. 203 (1959).

29. 503 F. 2d 625 (9th Cir. 1974) cert. denied 423 U.S. 875 (1975).

30. (1979) 5 ALRB No. 40.

31. (1983) 9 ALRB No. 44.

32. (1977) 3 ALRB No. 82, modified, McAnally Enterprises v. ALRB (February 24, 1984) 4 Civ. No. 19624, modified (1985) 11 ALRB No. 2 (modifying the Board's remedy).

33. (1978) 4 ALRB No. 62.

34. (1981) 7 ALRB No. 18.

35. (1980) 6 ALRB No. 44.

36. M. Caratan, Inc. (1978) 4 ALRB No. 83.

37. (1978) 4 ALRB No. 81.

38. (1983) 9 ALRB No. 69.

39. (1984) 10 ALRB No. 10.

40. Id.

41. Pioneer Nursery (1984) 10 ALRB No. 30.

42. (1983) 9 ALRB No. 31.

43. Vista Verde Farms v. ALRB, 29 Cal. 3d 307, (1981).

44. Cal. Lab. Code §1140.4(c) (West Supp. 1987).

45. (1977) 3 ALRB No. 67.

46. Martori Bros. Distributors v. ALRB, 29 Cal. 3d 721 (1981); Wright Line, Inc. (1980) 251 N.L.R.B. 150.

47. (1981) 7 ALRB No. 18.

48. 388 U.S. 26 (1967).

49. Bacchus Farms (1978) 4 ALRB No. 26.

50. Abatti Farms and Abatti Produce (1984) 10 ALRB No. 40.

51. Id.

52. (1978) 4 ALRB No. 26.

53. Cal. Lab. Code §1155.2(a) (West Supp. 1987).

54. Sometimes an employer refuses to bargain in order to test the certification of the union. By engaging in this "technical refusal to bargain," the employer can obtain court review of the Board's certification of the election. This process is necessary, since the ALRA and the NLRA contain no procedure for judicial review of election cases. (Cal. Lab. Code §1160.8). In the court proceeding, the employer

defends its bad faith bargaining by alleging that the union was not properly certified and is therefore not the elected representative of the employees.

55. NLRB v. Herman Sausage, Inc., 275 F. 2d 229, 232 (5th Cir. 1960).

56. Good discussions on surface bargaining and hard bargaining can be found in Montebello Rose Co. and Mount Arbor Nurseries (1979) 5 ALRB No. 64, aff'd, 119 Cal. App. 3d 1 (5th dist. 1981); Carl Joseph Maggio v. ALRB, 154 Cal. App. 3d 40 (4th dist. 1984); NLRB v. General Electric, 418 F. 2d 736. (2d Cir. 1969), cert. denied 397 U.S. 965, enforcing 160 NLRB 190.

57. (1979) 5 ALRB No. 64, aff'd, 119 Cal. App. 3d 1 (5th dist., 1981).

58. Cal. Lab. Code §1155.2(a) (West Supp. 1987).

59. NLRB v. Wooster Div. of Borg-Warner Corp., 356 U.S. 342 (1958).

60. 379 U.S. 203 (1964).

61. First National Maintenance Corporation v. NLRB, 452 U.S. 666 (1981).

62. Textile Workers v. Darlington Co., 380 U.S. 263 (1965).

63. Id.

64. (1983) 9 ALRB No. 49.

65. (1982) 8 ALRB No. 72.

66. See also Lu-Ette Farms, Inc. (1982) 8 ALRB No. 91, which reached the same result where the company began using a melon harvesting machine to move melons from pickers to the truck.

67. (1981) 7 ALRB No. 37.

68. (1983) 9 ALRB No. 61.

69. 159 Cal. App. 3d 758 (4th dist. 1984).

70. (1984) 10 ALRB No. 3.

71. (1984) 10 ALRB No. 49.

72. Pik'd Rite, Inc. and Cal-Lina, Inc. (1983) 9 ALRB No. 39.

73. (1979) 5 ALRB No. 64, aff'd, 119 Cal. App. 3d 1 (5th dist. 1981).

74. See also Sam Andrews' Sons (1983) 9 ALRB No. 24, where significant issues had not been discussed, few meetings had occurred on the disputed issues and the union had not assumed an uncompromising attitude on the key issue.

75. NLRB v. Reed & Prince Mfg. Co., 205 F. 2d 131 (1st Cir. 1953).

76. Nish Noroian (1982) 8 ALRB No. 25.

77. Tri-Fannucchi Farms (1986) 12 ALRB No. 8.

78. (1977) 3 ALRB No. 57.

79. (1980) 6 ALRB No. 58.

80. Other types of union security clauses are the closed shop (requires the employer to hire only union members), the preferential shop (requires the employer to give preference to union members when hiring), and maintenance of membership (requires employees to maintain their union membership throughout the collective bargaining agreement if they join).

81. A Federal District Court in 1985 ruled that the good-standing clause of §1153(c) was unconstitutional as a violation of a worker's first amendment rights

to free speech and association, Beltran v. State of California 617 F. Supp. 948 (S. D. Cal. 1985). This case is currently being appealed to the Ninth Circuit Court of Appeal.

82. Cal. Lab. Code §1153(c) (West Supp. 1987).

83. (1980) 6 ALRB No. 16.

84. Abood v. Detroit Board of Education, 431 U.S. 209 (1977); Machinists v. Street, 361 U.S. 807 (1960).

85. Ellis v. Brotherhood of Railway Clerks, 466 U.S. 435 (1984).

86. UFW and Sun Harvest, Inc. (1983) 9 ALRB No. 40; UFW (Scarbrough) (1982) 8 ALRB No. 103.

87. 156 Cal. App. 3d 312 (1st dist. 1984).

88. (1986) 12 ALRB No. 16.

89. Tex-Cal Land Management, Inc. v. ALRB, 24 Cal. 3d 335 (1979).

90. M. B. Zaninovich (1984) 9 ALRB No. 63.

91. See, for example, Giumarra Vineyards (1981) 7 ALRB No. 24.

92. Phelps Dodge Corp v. NLRB, 313 U.S. 177 (1941).

93. Abatti Farms, Inc., Abatti Produce, Inc. (1983) 9 ALRB No. 59.

94. Lu-Ette Farms, Inc. (1982) 8 ALRB No. 55.

95. (1978) 4 ALRB No. 36.

96. Abatti Farms, Inc. and Abatti Produce, Inc. (1983) 9 ALRB No. 59.

97. NLRB Chairman McCulloch as cited in Adam Dairy (1978) 4 ALRB No. 24.

98. At least one federal appeals court has held that the NLRB has the power to order the remedy (International Union of Electrical, Radio and Machine Workers, AFL-CIO v. NLRB (Tiidee Products), 426 F. 2d 1243 (DC Cir. 1970), rehearing denied 431 F. 2d 1206, cert. denied, 75 LRRM 2752 (1980)). The Board, however, has concluded that it does not have proper authority (Ex-Cell-O (1970) 186 NLRB No. 20) based on a Supreme Court ruling that the NLRB cannot compel agreements on a particular provision (H. K. Porter Co. v. NLRB, 397 U.S. 99 (1970)).

99. Cal. Lab. Code §1160 (West Supp. 1987).

100. (1978) 4 ALRB No. 24.

101. (1978) 4 ALRB No. 25, rev'd on other grounds, 86 Cal. App. 3d 448 (3d dist. 1978).

102. Id. at 10.

103. 26 Cal. 3d 1 (1979).

104. J. R. Norton (1980) 6 ALRB No. 26.

105. Dal Porto and Sons v. ALRB, 191 Cal. App. 3d 1195 (3d dist. 1987).

106. Id. at 1207.

107. UFW (Maggio) (1986) 12 ALRB No. 16.

108. Adam Dairy (1978) 4 ALRB No. 24.

109. (1978) 4 ALRB No. 39, rev'd and remanded on other grounds, 26 Cal. 3d 1 (1979).

110. Holtville Farms, Inc. (1984) 10 ALRB No. 13; Kyutoku Nursery (1982) 8 ALRB No. 73.

111. Robert H. Hickam (1983) 9 ALRB No. 6.

112. (1984) 10 ALRB No. 42.

113. Id. at 20.

114. (1983) 9 ALRB No. 74, rev'd and remanded on the issue of bad-faith bargaining 3/11/86.

Chapter 7

Strikes, Boycotts and Pickets

A. STRIKES

The right to strike is a principle now firmly embodied in public policy, but in the past, U.S. policy and law have opposed and even prohibited strikes. From the concerted work stoppages in the late eighteenth century until 1842, the courts prohibited strikes by declaring them to be criminal conspiracies of workers. This legal doctrine held that unions were unlawful combinations of workers trying unlawfully to raise the price of labor and thus take undue advantage of the public. Most juries found union activities illegal, in part because jurors were required to own property, so that only merchants and other employers served as jurors. Unions were not considered lawful until 1842, when the Massachusetts Supreme Court held in the case of *Commonwealth v. Hunt* that unions in themselves were not unlawful and that union members could not be jailed for union activities unless union objectives or the means used to achieve these objectives were unlawful.

In the late 1880s courts again prohibited strikes, this time by granting injunctions to employers who requested them. Injunctions are court orders directing a person or group to refrain from certain actions such as strikes, boycotts, or picketing, and are issued by a judge acting alone. Employers petitioned courts for injunctions against labor union activity including strikes, and the judges, reflecting contemporary economic and social attitudes, acted as legislators by deciding to prohibit strikes and other union conduct. Blanket injunctions were issued, culminating in a 1911 injunction which forbade anyone from speaking or writing to further labor union activity.

The era of blanket injunctions began drawing to a close with passage of the Clayton Antitrust Act in 1914, which prohibited injunctions against unions unless irreparable injury was shown or there was no alternative remedy. Courts could no longer prohibit peaceful picketing, strikes, or boycotts. The Norris-La Guardia Act of 1932 further restricted the ability of courts to issue injunctions in labor disputes. The National Labor Relation Act of 1935 explicitly gave workers the right to engage in concerted work stoppages and to retain their employee status while striking.

Today, strikes are one of organized labor's main economic weapons.[1] Strikes are an important part of the collective bargaining process, making "the risk of loss so great that compromise is cheaper..."[2] Both the NLRA and the ALRA protect the right of workers to engage in strikes, but certain types of strike activity are limited by one or both acts.

1. Lawful Strike Activity

The most common type of strike occurs when there is a wage dispute as a contract is being negotiated for the first time or renegotiated; e.g., workers demand a 6 percent raise, the employer refuses to offer more than 3 percent, and the workers strike to pressure the employer. This is known as a *economic strike* because the central issue is the economic terms of the contract. It is also described as a *primary* strike, since it is a dispute between the union and the primary (direct) employer.

In most industries, the economic strike has become an accepted part of collective bargaining. Most employers cease operations during the strikes, thus avoiding the violence sometimes generated by strikebreakers. Most economic strikes have not been fights by the union or employer for survival, although the 1980s recession, deregulation, and anti-union attitudes have made bitter strikes more common.

Agricultural labor relations have not matured to the same stage as industrial relations and farmworker strikes have often been major battles. Violence is not uncommon and is sometimes encouraged by employers who hire security guards brandishing firearms, use attack dogs in the fields, and import strikebreakers. Unions also sometimes engage in picket line violence against strikebreaking workers or sabotage employer property. One reason for the strong response of farm employers is their fear of losing their entire crop during a harvest season strike;[3] workers fear loss of their seasonal earnings.

A second type of strike is the *unfair labor practice strike*, which is initiated in response to an employer's unfair labor practice. The cause may be an employer's bad faith bargaining, the discharge of a worker for engaging in protected activities, or another unfair labor practice, and the workers stop working to protest the employer's unlawful activity.

Both economic strikes and unfair labor practice strikes are protected worker activities, so an employer may not fire workers for striking, but the employer can hire replacement workers in order to continue operations. Under the NLRA, an employer may hire permanent workers to replace economic strikers and then refuse to rehire strikers who have been permanently replaced after the strike is settled.

However, the strikers retain a right to preferential hiring if a vacancy occurs or if the permanent worker quits. A temporary replacement worker, on the other hand, can be bumped by a returning striker. Unfair labor practice strikers cannot be permanently replaced and they are entitled to reinstatement to their former jobs when they return to work.

In agriculture, the principle of "permanent replacements" becomes fuzzy because of normal work force turnover and the seasonal nature of the work. Faced with the issue of reinstatement rights for economic strikers, the ALRB has found that informality in the hiring process reflects a general lack of continuity in agricultural employment:

> Fluctuating labor needs have resulted in informal hiring practices. Some employers hire and lay off workers in accordance with weather or market conditions; such changes in employment may take place rapidly. Workers are often employed on a first-come, first-serve daily basis by appearing at a customary pick-up point at the start of the work day. Hiring at the boundary of an employer's field also occurs. Moreover, labor contractors are frequently used to supply growers with labor or to supplement a regular labor force, and farmworkers often work for a labor contractor who provides workers to several growers within one season.[4]

The ALRB concluded that, given such employment practices, an employer interferes with workers' rights to organize and strike in violation of 1153(c) if he refuses to rehire strikers at the conclusion of the strike, unless he or she *had* to hire permanent replacements to continue operations. The ALRB will accept the employer's characterization of its replacement workers as "permanent" for the current season, so grape workers going on strike in September 1987 need not be rehired to pick grapes in 1987 if they request reinstatement this year. However, since there is usually a large turnover in the work force by the next season, the ALRB requires the employer to rehire all returning strikers next year (September 1988). Employers are granted an exception to this rule if they show that it was necessary to offer permanent employment which would continue to the following season in order to recruit replacement workers during the strike.

One justification for not rehiring strikers is a good faith belief by the employer that strikers have engaged in serious strike misconduct. The employer must demonstrate a good faith belief that strikers who are not rehired committed substantial misconduct. Then the burden shifts to the ALRB General Counsel to show that misconduct did not occur. If the General Counsel proves that the striker did not engage in misconduct, the striker must be rehired.[5]

Two more types of strikes are lawful within certain limits under the NLRA— recognitional strikes and jurisdictional strikes. A *recognitional strike* occurs when the union seeks recognition as the representative and collective bargaining agent of employees. The NLRA allows a union to strike and picket a primary employer for 30 days if no other union has been certified, the employer has not recognized any other union, and there is no election bar.[6] Under the ALRA, a recognitional strike by an uncertified union would be for an unlawful purpose, since an employer can only recognize a union which has been certified by the ALRB pursuant to a secret ballot election or bargaining order. It is a union unfair labor practice under both the NLRA and ALRA for a union to force an employer to recognize and bargain with it if another union has been certified as the bargaining representative for workers.[7]

A *jurisdictional strike* is one in which two unions have collective bargaining agreements in different bargaining units with the same employer and each union claims its workers are entitled to a certain work assignment. A jurisdictional strike is unlawful under the NLRA only if the employer assigns work to the "wrong" union, i.e., a union which was not certified to do that type of work. In all other situations jurisdictional strikes are prohibited,[8] in order to protect a neutral employer from being victimized by two rival unions. Under the ALRB, jurisdictional strikes are not possible since all agricultural employees are in the same bargaining unit.

2. Unlawful Strike Activity

Violence and threats of violence accompanying a strike are unlawful and can be enjoined (halted) by the courts. Violent conduct coerces nonstriking employees and therefore is a union unfair labor practice under ALRA section 1154(a)(1). After an employer files an unfair labor practice charge against a union for strike violence, the ALRB or the employer can petition the local superior court for injunctive relief. Before a union can be held responsible for strike violence, however, the evidence must show that the violence was the result of union policy or was encouraged or engaged in by union officials.

Secondary strikes are unlawful under both the ALRA[9] and NLRA.[10] A *secondary strike* is one in which the objective is to force an employer to stop using, selling, or transporting the products of another company. An example would be if union X called a strike against winery Y to force it to stop buying grapes from a nonunion grower Z. This form of pressure on a neutral employer is undesirable, since the secondary employer is the victim of the dispute between the union and the primary employer and it has no power to resolve the primary dispute.

Intermittent strikes are an unlawful series of work stoppages. *Partial strikes* refer to partial work stoppages or work slowdowns. Intermittent and partial strikes are unprotected activity because they prevent an employer from continuing the business with replacement workers.[11] However, a one-time work stoppage to protest wages or working conditions is a protected concerted activity. Two[12] and perhaps three work stoppages may be protected, so the exact line between protected concerted activity and unprotected intermittent strikes is sometimes difficult to discern. It is clear, however, that a union sponsored series of weekly strikes would be unlawful.

A strike which has an objective prohibited by the ALRA or any other statute is illegal. Secondary strikes and certain recognitional and jurisdictional strikes fall into this category. Other unlawful objectives include (a) forcing an employer to sign an agreement not to do business with another producer, processor, or manufacturer (a "hot cargo" agreement) and (b) forcing an employer to join a labor or employer organization.[13] An example of an unlawful hot cargo agreement between union X and winery Y is a clause in the contract saying that the winery would not buy grapes from grower Z unless grower Z has a contract with union X.

B. CONSUMER BOYCOTTS

Consumer boycotts of a good produced by a primary employer can take two forms: (1) the union can try to persuade consumers not to purchase the product or (2) the union can try to persuade consumers not to patronize stores that sell the product. An example of a product boycott strategy is if union X won an election at winery Y, then asked consumers at market Z not to buy Y's wine in order to pressure the winery to agree to the union's wage demands. An example of the store boycott strategy is if union X won an election at winery Y and asked consumers not to shop at market Z because it handles winery Y's wines. The second type of boycott is considered a secondary boycott, since it involves action against market Z, which is not involved in the primary dispute.

Both types of boycotts are lawful under the ALRA, although the secondary boycott is lawful only if the union is certified as the representative of the primary employer's employees and does not cause a secondary strike. For example, it is lawful for the union to urge consumers to boycott the store because it is selling winery Y's wine, but it is not lawful for the union to persuade store clerks to go out on strike to protest the store's handling of a certain wine.

Secondary boycotts were made illegal under the NLRA by the 1947 Taft-Hartley amendments.[14] Opponents of the secondary boycott argued that it should not be allowed under the ALRA; proponents argue that the secondary boycott was legal under the NLRA from 1935 to 1947 and it should be permitted during the early years of agricultural labor relations.

C. PICKETING

Picketing is a common means for unions to advertise a labor dispute. Since picketing is a form of speech, it is protected by the First Amendment of the U.S. Constitution as long as it does not involve unlawful actions.

Picketing a primary employer in the course of a lawful strike is legal unless it involves violence, threats of violence, or the blocking of entrances to the employer's facility. In these cases, the picketing can be enjoined by state courts at the request of either the ALRB or the employer, after the employer has filed an unfair labor practice charge with the ALRB.[15] Picketing to inform consumers of a lawful boycott of a product or that the store is selling the boycotted product is legal under the ALRA as long as entrances are not blocked.[16] Under the NLRA, picketing to accompany a secondary boycott is unlawful, since such a boycott is unlawful.

There are three additional forms of picketing. *Recognitional picketing* is picketing with the objective of forcing an employer to recognize the union as the collective bargaining agent of the employees. Recognitional picketing is prohibited by the ALRA and is also prohibited by the NLRA if another union has been certified or if a valid election has been held within the past 12 months.[17] Under the NLRA, a union can picket for recognition for up to 30 days if no other union is certified and there is no election bar.[18]

An *informational picket line* is one that informs the public of a dispute or point of view. Both the ALRA and NLRA protect informational picketing for the purpose of advising the public that an employer does not employ members of a labor organization or does not have a union contract. However, the picketing must not result in a strike against the employer or a refusal by employees of another company to pick up, deliver, or transport its goods.[19]

Residential picketing is picketing by a union at workers' homes with the intent of encouraging them to join a strike. The ALRB has ruled that picketing outside employees' homes in an intimidating coercive manner is an unfair labor practice, since the coercive impact is greater at a person's home.[20] However, residential picketing is allowed outside strikebreakers' homes when it is tightly controlled. In *California Coastal Farms vs. Agricultural Labor Relations Board*,[21] residential picketing was allowed when no more than two pickets were located on a single block; the picketing did not begin before 9:00 a.m. or extend past sunset; no more than two pickets approached a house to talk to the strikebreakers; and the pickets did not make verbal threats.

D. SUMMARY

Lawful strikes are a legitimate exercise of union power. There are two kinds of strikes: economic strikes and unfair labor practice strikes. The difference between these strikes is significant under the ALRA; ULP strikers are entitled to their jobs back when they decide to unconditionally return to work, but economic strikers who have been replaced may have to wait a year before they are entitled to their old jobs.

Strike violence occurs because both workers and employers face significant losses from a long strike. Violence is a violation of ALRA sections 1153(a) and 1154(a)(1). Secondary strikes are also prohibited by the ALRA, and intermittent and partial strikes are not protected activities under ALRA section 1153(a).

Primary and secondary boycotts are lawful under the ALRA, and have been used very successfully by farmworker unions. Informational picketing is lawful unless it creates a secondary strike. Residential picketing is unlawful if it intimidates workers in their homes.

NOTES

1. Other sources of power used by unions include boycotts, political action and restriction of the labor supply.

2. Morris, C. J., The Developing Labor Law, Washington, D.C.: The Bureau of National Affairs, Inc., 1971, p. 517.

3. One study has shown, however, that total grower revenues and profits actually rose during a 1979-80 lettuce strike. Carter, C.A. et al., "Labor Strikes and the Price of Lettuce," Western Journal of Agricultural Economics 6(1981):1-14.

4. Seabreeze Berry Farm (1981) 7 ALRB No. 40 at 5-6.

5. Bertuccio Farms (1984) 10 ALRB No. 52

6. National Labor Relations Act §8(b)(7), 29 U.S.C. §158(b)(7) (1982).

7. National Labor Relations Act §8(b)(7)(A), 29 U.S.C. §158(b)(7) (1982); Cal. Labor Code §1154(d)(3) (West Supp. 1987).

8. National Labor Relations Act §8(b)(4)(D), 29 U.S.C. §158(b)(4)(D) (1982).

9. Cal. Lab. Code S1154(d).

10. National Labor Relations Act §8(b)(4)(B) 29 U.S.C. §158(b)(4)(B) (1982)..

11. Sam Andrews' Son, (1983) 9 ALRB No. 24.

12. Bertuccio Farms (1984) 10 ALRB No. 52.

13. Cal. Lab. Code §1154(d)(1) and §1154.5 (West Supp. 1987).

14. National Labor Relations Act §8(b)(4), 29 U.S.C. §158(b)(4) (1982).

15. Bertuccio vs. Superior Court 118 Cal. App. 3d 363 (1981).

16. Cal. Lab. Code §1154(d)(4) (West Supp. 1987).

17. Cal. Lab. Code §1154(g) (West Supp. 1987); National Labor Relations Act §8(b)(7), 29 U.S.C. §158(b)(7)(C) (1982).

18. National Labor Relations Act §8(b)(7)(c).

19. Cal. Lab. Code §1154(g) (West Supp. 1987); National Labor Relations Act §8(b)(7)(c).

20. United Farm Workers of America, AFL-CIO (Marcel Jojola) (1980) 6 ALRB No. 58.

21. 111 Cal. App. 3d 734 (1st dist. 1980).

Chapter 8

The Future of Farmworker Collective Bargaining

The United Farm Workers (UFW) union was 25 years old in 1987, and it remains the dominant farmworker union in the United States. Five other California farmworker unions also represent farmworkers, but none approaches the UFW in stature as a spokesperson for farmworkers.

Farmworker union activity in other states is also limited in the mid-1980s. The Arizona Farmworkers Union claims 15,000 members, but represents only about 1,000 legal Mexican farmworkers working temporarily in the United States. The Farm Labor Organizing Committee represents 1,500 tomato and cucumber harvesters on Ohio and Michigan farms that grow these commodities for Campbells and Heinz, and the International Longshoremen's Union represents farmworkers in Hawaii. There is farmworker protest or union activity in Florida, Washington, Texas, and New Jersey, but few contracts. California is the only major agricultural state with an ALRA and ALRB; in states such as Florida, Washington, and Texas, there is no law or administrative agency to regulate the relationships between workers, unions, and employers.

Many farm labor reformers predicted that the ALRA and ALRB would bring peace and stability to relations between California farmers and farmworker unions. They were wrong. Farmers initially tried to amend the ALRA so that it conformed to the NLRA, and the ALRB was first damned by growers and then by the UFW. After 12 years of ALRB activity, many employers and unions are still unwilling to resolve their differences through collective bargaining or to accept the decisions of the ALRB.

The UFW has been losing members and contracts in the 1980s for a variety of reasons. The proliferation of farm labor contractors (FLCs) who employ hard-to-organize illegal alien workers provides formidable competition for the UFW because FLCs can supply a lower-cost alternative workforce to nonunion employers. For example, the Ventura (Ca.) lemon harvest switched from a largely

UFW-represented workforce in the late 1970s to a nonunion FLC-supplied workforce by the mid-1980s.[1] A Governor elected in 1982 appointed persons to the ALRB who appear less interested in vigorously pursuing worker charges that employers violated their ALRA rights; the UFW has concluded that the ALRB has lost its usefulness and has called for elimination of the agency's funding.[2] Finally, the UFW has suffered from a variety of leadership crises; as long-time UFW supporters left, they were often replaced by close relatives of founder Cesar Chavez.

In mid-1987, an Imperial county judge assessed a $1.7 million fine against the UFW for encouraging violence during a vegetable strike in 1979-80. The UFW is appealing the decision, charging inter alia that the judge was biased and that having to post a bond in order to appeal the decision would bankrupt the union; however, the UFW posted the bond to appeal the decision and continued its grape boycott activities.

As the UFW grapples with its internal problems, the other California farmworker unions remain relatively small and limited to certain commodities and areas. Farmworker unions and collective bargaining generate much less interest in the mid-1980s than they did just a decade ago. Thus, an examination of underlying union strength and the future of the ALRA and ALRB is timely.

A. UNION POWER

Unions have four sources of power: strikes, political action, boycotts, and control over the supply of labor.[3] The traditional oversupply of immigrant farmworkers without job options makes it hard for a farmworker union to control the supply of labor. These immigrant workers often become strikebreakers, further eroding a union's bargaining leverage.

A farmworker union able to prevent strikebreaking can still find that strikes yield pyrrhic victories because most California fruits and vegetables have inelastic consumer demands and are overproduced. An inelastic demand means that consumers buy about the same amount of lettuce even if its price rises, and overproduction means that only two-thirds of the available lettuce is typically harvested because lettuce prices do not justify hand-harvesting a field 4 or 5 times to get all the lettuce. Overproduction means that partially successful strikes can increase grower prices and revenues. The 1979 UFW lettuce strike, for example, helped to increase grower prices from $7 to $21 per carton and double strike period lettuce revenues from $35 million to $70 million.[4] These extra profits were unevenly distributed (the major beneficiary was Bud Antle, which had a Teamster's contract and thus was unaffected by the strike), but inelasticity and overproduction suggest that most growers might benefit from partial strikes with strike insurance that pools and distributes excess profits, as some airlines did before deregulation. Effective farmworker strikes must drastically curtail production before they hurt employers, and such strikes are difficult to mount because there are several competing farmworker unions; because growers can usually find strikebreaking workers; because there are many nonunion producers of fruits and vegetables in California and in other states; and because imports increase when fruit and vegetable prices rise.

Since strikes and labor supply controls have limited effects, farmworker unions tend to rely on political activities and consumer boycotts to persuade farmers to agree to union demands. It is these nontraditional union weapons which generated publicity and power for the UFW in the 1970s, so the UFW rationally devotes considerable efforts to "nonfarm" politics and boycott activities. Politics won for farmworker unions the special features of the Agricultural Labor Relations Act: quick elections, ALRA good standing, makewhole penalties, and more permissible secondary union activities. These legal tools and the boycott vulnerability of corporate farms help to explain why the UFW has been most successful in corporate California agriculture.

Corporate subsidiaries are most vulnerable to consumer boycott activities because nonfarm corporations often bought farmland in the inflationary 1970s in anticipation of land price inflation and to supply a branded product at premium prices. These corporations often dealt with unions in their nonfarm businesses, and were more receptive to union organizers than more traditional farmers. Indeed, the UFW's first major success came in 1965 in the vineyards of Schenley Industries, a liquor business worried about a consumer boycott of its products. During the early 1980s Amfac, a corporate farmer and owner of department stores and hotels, was vulnerable to threats that "poor farmworkers" would picket its downtown stores during the major shopping season if Amfac did not agree to UFW demands.

Farmland prices fell in the 1980s, and many corporate farms operated by nonfarm corporations were sold after their owners experienced the extreme price and profit fluctuations that are common in agriculture. The branded product strategy was also a mixed blessing: the brand identification that generated premium prices also expedited union-inspired consumer boycotts. Bruce Church, for example, developed a premium Red Coach lettuce label that facilitated a UFW boycott against its lettuce. Integrated grower-packers who successfully develop brand identification also help unions to mount targeted product boycotts more effectively.

Branded fruits and vegetables may continue to aid unions. Some corporate farms have avoided branded commodities because they recognize their boycott vulnerability, but Castle and Cooke and Campbell's are corporations which are expanding their labor-intensive production and packaging operations, increasing the potential boycott leverage of farmworker unions. Thus, the corporate structure of California agriculture helps and hurts farmworker unions. Unions are helped by the corporate willingness to deal with unions and corporate vulnerability to boycott pressures. But corporations can also shift their investments out of agriculture; the

UFW's breakthrough vegetable contract in the early 1980s was with Sun Harvest, a United Brands subsidiary which went out of business in 1983. Vulnerable corporations have adopted several tactics to fend-off farmworker unions, including producing fruits and vegetables abroad or converting themselves into middlemen or processors who contract with "independent farmers" to grow crops for them.

Another long-run threat to farmworker unions is the shifting of production from the union-dominated coastal valleys to the largely nonunion San Joaquin Valley of California. The Salinas and Ventura coastal areas still account for most of the UFW's field worker contracts, but differences in land prices, new plant varieties that can thrive in the Central Valley, and wage differences are encouraging more fruit and vegetable growers to shift to the largely nonunion San Joaquin Valley. Farmworker unions are also preparing to make this geographic transition, and in 1984-85 the UFW announced a major organizing drive in the San Joaquin Valley citrus industry.

Corporate farming facilitates union organizing, but even if farmworker unions successfully follow corporate growers from the coastal valleys to the San Joaquin Valley, they must still grapple with labor-displacing mechanization. Mechanization can affect unions in an obvious way, by substituting machine operators and sorters for handharvest workers, or by changing the way work is done so that growers can employ new groups of workers. Most of California's arduous hand-harvesting is done by young men who are motivated to work hard by piece rate wages. A mechanical aid such as a conveyor belt that eliminates some of the heavy lifting and carrying allows employers to substitute women for men and to switch to an hourly wage system because the employer can control the speed of the machine through the field. Micro-technologies that permit sizing equipment to be put on field machines and portable cooling devices mean that some commodities which are now picked by crews of young men for piece rate wages and hauled to a packingshed can be picked and packed by crews of diverse workers in the field.

Mechanization and mechanical aids promise to have opposite effects on union bargaining power. Labor-saving mechanization threatens to displace farmworkers and undercut unions, but mechanical aids increase the number of farm jobs by moving nonfarm packinghouse jobs into the field. The number of "farmworkers" may increase if mechanical aids move nonfarm packing jobs into the fields faster than mechanization eliminates farm jobs. The number of packingshed jobs, on the other hand, would decrease.

Fieldpacking promises to change the composition of the farm work force and may strain relationships between nonfarm packingshed unions and farmworker unions. The commodities most affected by fieldpacking, vegetables and melons, have traditionally been hauled from the field into a union-organized packingshed to be prepared for market. Packingshed workers are usually represented by a union with few farmworker members, such as the Teamsters or FFVW, but fieldworkers are represented by the UFW or the IUAW. As packing operations move into the fields, packingshed unions lose members, so fieldpacking may increase inter-union rivalry if packingshed unions attempt to follow their members into the fields.

Unions are being buffeted by trends which increase (fieldpacking) and decrease (mechanization) their bargaining power, but grower bargaining power is also changing. The bargaining power of farmers is their ability to live with or stay in business despite union-called strikes or boycotts. In nonfarm industries, the strike is a powerful weapon because, even if management wanted to replace 500 or 1,000 manufacturing workers, it is not easy to find replacement workers quickly who can operate specialized manufacturing equipment. However, in California agriculture, farmers have traditionally been able to replace striking workers with immigrant workers, and the many entrances and exits on most farms often make it hard to notify strikebreaking immigrants of the strike.

Farmers learned to cope with strikes by assuring themselves an ample supply of vulnerable immigrant workers. However, the Immigration Reform and Control Act (IRCA) of 1986 may reduce the availability of illegal alien farmworkers. If farmers must turn to American and legal temporary foreign workers, farmworker unions may gain more bargaining leverage by lobbying the federal government to minimize the importation of foreign farmworkers.

Growers demonstrated their political muscle by winning a guestworker program under IRCA, but they have been less successful in coping with union-inspired political and boycott activities. The UFW has been the primary innovator in both areas, amassing significant funds for a political action committee (PAC) and generating considerable support from churches and local politicians such as mayors and city councils in UFW boycotts. However, farmers have also established PACs and have begun to wage aggressive counter-boycott publicity campaigns.

B. THE ECONOMIC IMPACTS OF FARMWORKER UNIONS

Unions seek to raise wages and improve working conditions. After a union is certified as the bargaining agent for a group of workers, it seeks to negotiate a collective bargaining agreement with the employer and tries to ensure that other employers offer similar wages and working conditions. In most economic models of union behavior, unions are confronted by a trade-off between higher wages and more employment.

Most economic models of unions imagine a single union which has been certified to represent most of the workers in an industry, such as a carpenters union which has organized the local carpenters. If unions have enough power to control the supply of labor, then the union can select any wage-employment trade-off it wants on the firm's demand for labor curve. Three points are illustrated in Figure 8.1; the union can maximize worker rent at r — rent is the extra wages workers get because the union controls the supply of labor. Alternatively, the union can

Figure 8.1

The Union Wage-Employment Tradeoff

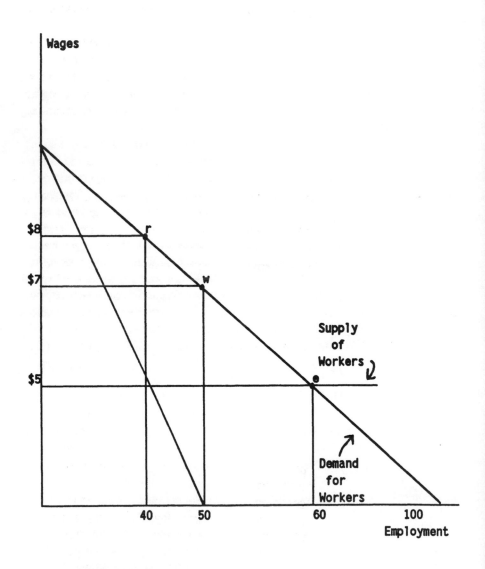

max rent (r) = 40 x $8 = $320
max total wages (w) = 50 x $7 = $350
max total employment (e) = 60 x $5 = $300

maximize employment at **e**, but at a lower wage. The third union option is to maximize the employer's wage bill at **w**. Thus, the typical economic model of an all-powerful union asserts that the demand for labor is negatively-sloped because consumers will buy more only at lower prices, and that the union can select any point along this demand curve to the left of **e** (the supply curve indicates that workers will not work for a wage below **e**).

Even all-powerful unions are comprised of people, and the negatively-sloped demand for labor forces a trade-off between wages and employment. Most unions elect their leaders, so union leaders must usually be accountable to members who are laid-off if the union pushes wages too high. To minimize this painful trade-off, many unions attempt to reduce the elasticity of demand for labor, or the extent to which an increased wage leads to layoffs.

The elasticity of demand for labor—the trade-off between higher wages and less employment—depends on the ease with which workers can be replaced by machines or chemicals; the elasticity of demand for the product produced (how much less do consumers buy when higher wages cause higher prices); the elasticity of supply of labor substitutes (if all farmers switch to tomato harvesting machines, does the price of tomato harvesting machines jump?); and the importance of labor in total production costs (if labor is 50 percent of total production costs, then a 10-cent wage increase means a five-cent increase in the product price). Unions attempt to reduce the elasticity of demand for labor by restricting mechanization, urging consumers to buy union-made products, and organizing all employers in an industry so that nonunion firms do not have lower wages and prices. The UFW once represented most of California's lettuce and grape workers, but in the mid-1980s it represents only a majority of the workforce in mushrooms. The UFW has been most successful in convincing Americans not to buy certain (nonunion) commodities, such as table grapes.

Farmworker unions did not usually have the power to select a particular wage-employment trade-off. Indeed, farmworker unions sometimes deal with groups of employers who, by acting collectively, exercise control over how many workers find employment. Instead of a powerful union making wage-employment trade-offs, powerful groups of employers have historically had monopsony power to determine farm wages.

Monopsony power means that employers have a demand for labor schedule indicating how many workers they will hire at various wages and a supply schedule which indicates how many workers are available to work at various wages. If employers can attract more workers only by offering a higher wage, then the supply curve is positively-sloped and has a Marginal Factor Cost (MFC) lying above it, reflecting the fact that the extra cost of adding a 10th worker to work requires that employers offer all workers $4 hourly (Figure 8.2). If an employer association is a monopsony (the only or major employer of farmworkers), then it will hire until $MFC = D_L$, or 30 workers at a wage of $6 hourly. The value of the 30 worker crew

122

Figure 8.2

Monopsony Employer Hiring

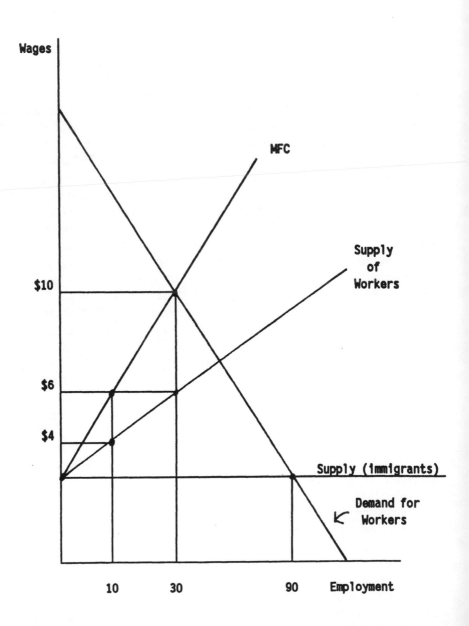

to the employer is $300 hourly, even though the employer is paying the crew only $180 hourly—this difference is called "monopsonistic exploitation."

Farmers have historically sought to make the supply of workers horizontal or elastic at a minimum wage. One way to assure themselves such a supply of workers is to get the government to admit immigrant workers if American workers demand more than a minimum wage to be farmworkers. There is a virtually unlimited supply of immigrant workers available at the U.S. minimum wage of $3.35 per hour in the 1980s, so that if farmers get the government to agree to admit immigrant workers to fill farm labor shortages, unions will be hard-pressed to raise farmworker wages.

A monopsonistic employer association must pay higher wages to expand employment; expanding employment from 9 to 10 workers requires that wages be raised from $3.78 to $4 hourly, so the marginal factor cost of the 10th worker is $6. The monopsony with this demand for workers will hire a crew of 30 workers at a wage of $6 hourly.

In most instances, neither farmers nor unions have complete labor market power. In these situations, the supply of and demand for farmworkers determines the framework for bargaining, but the exact wage-employment combination that is reached depends on the skills and behavior of the negotiators. In agriculture, it is often hard to predict this wage-employment combination because bargaining is new and the union's major weapons—politics and boycotts—have effects which are hard to quantify. Farmers may not feel threatened by a boycott, and may "bargain hard" because they believe that it will take the union a long time to mount an effective boycott against them.

Farmworker unions have changed employment practices on some California farms. Unions typically reduce worker turnover by giving dissatisfied workers a "voice" to complain to management. In nonunion firms, by contrast, management learns of worker dissatisfaction more from "exits" of dissatisfied workers.[5] Unionization and seniority recall appear to have reduced turnover among seasonal workers on some California farms, especially in nurseries which offer less arduous and more year-round jobs. Farmworker union elections are often the only opportunity that illegal alien workers have to vote in the United States, since the ALRA protects workers regardless of their legal status.

There have been a few attempts to estimate the economic effects of the UFW on farmers, workers, and consumers, but most such analyses are biased. For example, a newspaper reported that:

Chavez is responsible for fundamental improvements in the lives of American farm workers. California rural laborers, who earned an average of $1.40 an hour in 1965 with no fringe benefits, earned an average of $4.51 an hour in 1980, with many union contracts providing $6 an hour plus health care, paid holidays and pensions.[6]

Such an accounting gives too much credit to the UFW by failing to credit the end of the Bracero program, revisions in labor laws, and other changes which influenced farm labor markets.

However, critics of the UFW also tend to attribute too much power to it. For example, one study estimated that the UFW and the ALRA "cost" society $76 million in increased costs and prices between 1974 and 1979,[7] but this study also attributed most of the changes occurring in agriculture to the UFW.

An unbiased analysis of the UFW's effects on California's farm labor market would indicate that the UFW has had significant but limited effects on farm wages and employment practices. On the largest fruit and vegetable farms, the UFW raised wages and introduced employment practices common in nonfarm markets, but the UFW and other unions have affected fewer than 1,000 California farms directly.

C. THE ALRA AND ALRB

The ALRA and ALRB have been a controversial experiment to "empower the powerless" and to bring peace and stability to relations between California farmworkers and growers. The UFW began the struggle for state-regulated collective bargaining with the mid-1960s grape boycotts; achieved more than it hoped for with the mid-1970s ALRA; and began to boycott the ALRB dispute resolution mechanism that it had struggled to achieve in the mid-1980s. The ALRA was based on the National Labor Relations Act and directed the ALRB to follow NLRB precedent. When the NLRB provided no precedent, as with the makewhole remedy, or when NLRB precedent was not applicable to agriculture, as with the access rule, the Board made policy decisions that had administrative and political ramifications. Many of these Board decisions have been appealed to reviewing courts, creating a backlog of unfair labor practice complaints.

The ALRA differs from the NLRA in election procedures and standards, access rules, and the makewhole remedy for bad-faith bargaining. In these areas the ALRB interpreted a sometimes ambiguous law with little or no NLRB precedent. The ALRB was not alone in interpreting the ALRA; appeals courts overturned part or all of about 20 percent of the ALRB decisions that were appealed to the courts. The ALRA and ALRB were also embroiled in politics: ALRB funding was halted for six months in 1976; the UFW developed a constitutional initiative in 1976 to guarantee funding for the ALRB; and legislative hearings were held to determine if the ALRA should be conformed to the NLRA. Since 1982, Governor Deukmejian has been replacing the ALRB members appointed by Governor Brown. In addition, the new Governor appointed a new General Counsel. The result has been that the ALRA has become a political football, with farmer and farmworker groups expending much money and effort to influence ALRB appointments.

Faced with a backlog of cases in certain areas such as makewhole remedies, the ALRB is faced with two major policy options. It can continue to litigate cases, some of which date from 1976, and hope that its remedies will eventually be enforced and workers will be compensated for unfair labor practices. This option is frustrated by the fact that the probability of locating farmworkers is reduced by every year of delay. In addition, when a court reverses an ALRB decision, the reversal may affect all cases pending on the same type of unfair labor practice, so the Board or a hearing officer must review previous conclusions in each pending case under the new standard, making it harder to eliminate the backlog.

A second option is to increase the use of settlements to resolve unfair labor practice claims. This option reduces the time between when the ULP was committed and the time that workers get compensated, increasing the probability that workers are found and paid. However, a quick settlement may undermine the goals of the ALRA by allowing an employer to retain some of the advantage it gained from the ULP because settlements usually involve less money for workers then they would get from a final enforcement.

A large backlog of cases, millions of dollars in potential liability, and political appointees who can enforce or rescind these liabilities guarantees that the ALRB will continue to be a political football. Both growers and unions have a great deal at stake in continuing to promote appointments to the ALRB and interpretations of the ALRA favorable to their interests.

What went wrong with the dream of a law and agency to bring peace and stability to labor relations in agriculture? From the UFW's perspective, the ALRB tilted against farmworkers and unions after a new Governor began appointing ALRB members in 1982. Farm employers, on the other hand, welcomed the end of an era of "bias" against them, and supported continued funding for the ALRB when the UFW tried to reduce or eliminate its funding in the mid-1980s. Most knowledgeable observers believe that the essentially one-industry, one-union ALRB was the victim of "too much politics," as governors successfully tried to appease their farmworker and grower constituents with their appointments. This perceived direct link between politics and ALRB appointments and decisions made the union or employer affected directly by an ALRB decision less likely to accept the decision and to learn to live with the law, and more likely to vow revenge in the next election.

The future of farmworker unions and collective bargaining is not clear. There are two short and extreme summaries of the future: many union proponents believe that "demography is destiny," while many employers are convinced that hard-to-organize immigrant guestworkers will keep collective bargaining the exception-to-the-rule in California agriculture. The "demography is destiny" argument notes that the Hispanics who dominate the field work force will also dominate the populations of many California towns and cities in coming decades,[8] helping to ensure that the political process is sensitive to farmworker concerns. The alternative view is that Mexican guestworkers will do much of the fieldwork in future years, and the urban Hispanics who vote in California elections will be more interested in education, housing, and transportation than in farmworker issues.

D. SUMMARY

The United Farm Workers union remains the dominant farmworker union in California and in the United States. However, the ALRA and ALRB have not been embraced by unions and growers as the neutral arbiters of inevitable work place conflicts, in part because both growers and unions have relied on political action as much as collective bargaining to achieve their goals.

Farmworker unions do not rely on labor's traditional weapons, the strike and the hiring hall, because political action and boycott activity have been far more effective. Farmworker unions rationally devote most of their energy to politics and boycotts, but the uncertainty of their impacts on growers has turned farmworker collective bargaining into competing public relations' campaigns.

The ALRA and ALRB are firmly entrenched as political footballs. As long as unions and employers believe that the ALRA and ALRB are vital to their interests, and that "their" politicians will modify the law or select appointees favorable to them, each side will devote its energies to getting its partisans into the agency. Thus, the future of the ALRA and ALRB is believed by both sides to rest in the hands of voters and politicians—the UFW believes that demography is destiny and that Hispanic voters will eventually force changes favorable to the union, while growers believe that a new wave of immigrant farmworkers will become more and more invisible to most Californians.

NOTES

1. Richard Mines and Philip Martin, Industrial Relations.

2. Harry Bernstein, "Farm Labor Board is Losing its Constituency," *Los Angeles Times*, January 12, 1986, Part V, p. 3, is typical of the calls to "defund" the agency that the UFW fought to create. The UFW argues that the share of (worker-filed) ULP charges dismissed by the ALRB rose to 80 percent in 1984-85 and that the ALRB collected only $290,000 from six growers in make-whole remedies since 1975, illustrating its irrelevance to farmworkers.

3. See Albert Rees. The Economics of Trade Unions (Chicago: University of Chicago Press, 1977).

4. Colin Carter et al., op cit.

5. Richard Freeman and James Medoff, What Do Unions Do? (New York: Basic Books, 1984).

6. "Cesar Chavez: Determined Shepard Gets Flak From UFW Flock," Washington Post, November 29, 1981, p. 2.

7. Rex Cottle et al., Labor and Property Rights in California Agriculture, (College Station: Texas A&M Press, 1982).

8. Hispanics are projected to double from 4.5 million or 19 percent of California's 1980 population to 9 million or 28 percent in 2000 and 16 million or 38 percent by 2030. Non-Hispanic Whites, by contrast, are projected to have a more stable population and a declining share: 16 million or 66 percent in 1980, 17 million or 52 percent in 2000, and 16 million or 38 percent in 2030. See L. Bouvier and P. Martin. Population Change and California's Future (Washington: Population Reference Bureau, 1985).

Appendix A

Selected Sections
of the ALRA

ALRA Section 1152

Employees shall have the right to self-organization, to form, join, or assist labor organizations, to bargain collectively through representatives of their own choosing, and to engage in other concerted activities for the purpose of collective bargaining or other mutual aid or protection, and shall also have the right to refrain from any or all of such activities except to the extent that such right may be affected by an agreement requiring membership in a labor organization as a condition of continued employment as authorized in subdivision (c) of Section 1153.

Section 1153

It shall be an unfair labor practice for an agricultural employer to do any of the following:

a. To interfere with, restrain, or coerce agricultural employees in the exercise of the rights guaranteed in Section 1152.

b. To dominate or interfere with the formation or administration of any labor organization or contribute financial or other support to it. However, subject to such rules and regulations as may be made and published by the board pursuant to Section 1144, an agricultural employer shall not be prohibited from permitting agricultural employees to confer with him during working hours without loss of time or pay.

c. By discrimination in regard to the hiring or tenure of employment, or any term or condition of employment, to encourage or discourage membership in any labor organization.

Nothing in this part, or in any other statute of this state, shall preclude an agricultural employer from making an agreement with a labor organization (not established, maintained, or assisted by any action defined in this section as an unfair labor practice) to require as a condition of employment, membership therein on or after the fifth day following the beginning of such employment, or the effective date of such agreement whichever is later, if such labor organization is the representative of the agricultural employees as provided in Section 1156 in the appropriate collective-bargaining unit covered by such agreement. No employee who has been required to pay dues to a labor organization by virtue of his employment as an agricultural worker during any calendar month, shall be required to pay dues to another labor organization by virtue of similar employment during such month. For purposes of this chapter, membership shall mean the satisfaction of all reasonable terms and conditions uniformly applicable to other members in good standing; provided, that such membership shall not be denied or terminated except in compliance with a constitution or bylaws which afford full and fair rights to speech, assembly, and equal voting and membership privileges for all members, and which contain adequate procedures to assure due process to members and applicants for membership.

   d. To discharge or otherwise discriminate against an agricultural employee because he has filed charges or given testimony under this part.

   e. To refuse to bargain collectively in good faith with labor organizations certified pursuant to the provisions of Chapter 5 (commencing with Section 1156) of this part.

   f. To recognize, bargain with, or sign a collective-bargaining agreement with any labor organization not certified pursuant to the provisions of this part.

Section 1154

   It shall be an unfair labor practice for a labor organization or its agents to do any of the following:

   a. To restrain or coerce:

   (1) Agricultural employees in the exercise of the rights guaranteed in Section 1152. This paragraph shall not impair the right of a labor organization to prescribe its own rules with respect to the acquisition or retention of membership therein.

   (2) An agricultural employer in the selection of his representatives for the purposes of collective bargaining or the adjustment of grievances.

   b. To cause or attempt to cause an agricultural employer to discriminate against an employee in violation of subdivision (c) of Section 1153, or to discriminate against an employee with respect to whom membership in such organization has been denied or terminated for reasons other than failure to satisfy the membership requirements specified in subdivision (c) of Section 1153.

c. To refuse to bargain collectively in good faith with an agricultural employer, provided it is the representative of his employees subject to the provisions of Chapter 5 (commencing with Section 1156) of this part.

d. To do either of the following: (i) to engage in, or to induce or encourage any individual employed by any person to engage in, a strike or a refusal in the course of his employment to use, manufacture, process, transport, or otherwise handle or work on any goods, articles, materials, or commodities, or to perform any services; or (ii) to threaten, coerce, or restrain any person; where in either case (i) or (ii) an object thereof is any of the following:

(1) Forcing or requiring any employer or self-employed person to join any labor or employer organization or to enter into any agreement which is prohibited by Section 1154.5.

(2) Forcing or requiring any person to cease using, selling, transporting, or otherwise dealing in the products of any other producer, processor, or manufacturer, or to cease doing business with any other person, or forcing or requiring any other employer to recognize or bargain with a labor organization as the representative of his employees unless such labor organization as the representative of such employees. Nothing contained in this paragraph shall be construed to make unlawful, where not otherwise unlawful, any primary strike or primary picketing.

(3) Forcing or requiring any employer to recognize or bargain with a particular labor organization as the representatives of his agricultural employees if another labor organization has been certified as the representative of such employees under the provisions of Chapter 5 (commencing with Section 1156) of this part.

(4) Forcing or requiring any employer to assign particular work to employees in a particular labor organization or in a particular trade, craft, or class, unless such employer is failing to conform to an order or certification of the board determining the bargaining representative for employees performing such work.

Nothing contained in this subdivision (d) shall be construed to prohibit publicity, including picketing for the purpose of truthfully advising the public, including consumers, that a product or products or ingredients thereof are produced by an agricultural employer with whom the labor organization has a primary dispute and are distributed by another employer, as long as such publicity does not have an effect of inducing any individual employed by any person other than the primary employer in the course of his employment to refuse to pick up, deliver, or transport any goods, or not to perform any services at the establishment of the employer engaged in such distribution, and as long as such publicity does not have the effect of requesting the public to cease patronizing such other employer.

However, publicity which includes picketing and has the effect of requesting the public to cease patronizing such other employer, shall be permitted only if the labor organization is currently certified as the representative of the primary employer's employees.

Further, publicity other than picketing, but including peaceful distribution of literature which has the effect of requesting the public to cease patronizing such other employer, shall be permitted only if the labor organization has not lost an election for the primary employer's employees within the preceding 12-month period, and no other labor organization is currently certified as the representative of the primary employer's employees.

Nothing contained in this subdivision (d) shall be construed to prohibit publicity, including picketing, which may not be prohibited under the United States Constitution or the California Constitution.

Nor shall anything in this subdivision (d) be construed to apply or be applicable to any labor organization in its representation of workers who are not agricultural employees. Any such labor organization shall continue to be governed in its intrastate activities for nonagricultural workers by Section 923 and applicable judicial precedents.

e. To require employees covered by an agreement authorized under subdivision (c) of Section 1153 the payment, as a condition precedent to becoming a member of such organization, of a fee in an amount which the board finds excessive or discriminatory under all circumstances. In making such a finding, the board shall consider, among other relevant factors, the practices and customs of labor organizations in the agriculture industry and the wage currently paid to the employees affected.

f. To cause or attempt to cause an agricultural employer to pay or deliver, or agree to pay or deliver, any money or other thing of value, in the nature of an exaction, for services which are not performed or not be performed.

g. To picket or cause to be picketed, or threaten to picket or cause to be picketed, any employer where an object thereof is either forcing or requiring an employer to recognize or bargain with a labor organization as the representative of his employees, or forcing or requiring the employees of an employer to accept or select such labor organization as their collective-bargaining representative, unless such labor organization is currently certified as the representative of such employees, in any of the following cases:

(1) Where the employer has lawfully recognized in accordance with this part any other labor organization and a question concerning representation may not appropriately be raised under Section 1156.3.

(2) Where within the preceding 12 months a valid election under Chapter 5 (commencing with Section 1156) of this part has been conducted.

Nothing in this subdivision shall be construed to prohibit any picketing or other publicity for the purpose of truthfully advising the public (including consumers) that an employer does not employ members of, or have a contract with, a labor organization, unless an effect of such picketing is to induce any individual employed by any other person in the course of his employment, not to pick up, deliver, or transport any goods or not to perform any services.

Nothing in this subdivision (g) shall be construed to permit any act which would otherwise be an unfair labor practice under this section.

h. To picket or cause to be picketed, or threaten to picket or cause to be picketed, any employer where an object thereof is either forcing or requiring an employer to recognize or bargain with the labor organization as a representative of his employees unless such labor organization is currently certified as the collective-bargaining representative of such employees.

i. Nothing contained in this section shall be construed to make unlawful a refusal by any person to enter upon the premises of any agricultural employer, other than his own employer, if the employees of such employer are engaged in a strike ratified or approved by a representative of such employees whom such employer is required to recognize under this part.

Appendix B

ALRB Cases

A. INTRODUCTION

The ALRA establishes election procedures for workers to select a collective bargaining representative or no union representative. The ALRA also enumerates unfair labor practices (ULPs) which employers and unions are prohibited from committing. These unfair labor practice provisions protect workers who are exercising their rights under the ALRA. If any person believes that an election was unfair because of coercive pre-election conduct or because technical election requirements were not met, he or she can file objections to the election with the ALRB. If someone believes that an ULP was committed, he or she can file an ULP charge with the ALRB. Procedures for handling election objectives and ULP charges differ, but each moves through several levels of ALRB and court review.

Election objections from workers, employers or unions are filed with the Executive Secretary (ES) of the ALRB. The ES reviews the objections and decides if they have merit. Election objections have merit if the cumulative effect of the facts alleged would, if proven true, constitute grounds for overturning the election and if the allegations are supported by affidavits. A hearing is held to resolve meritorious objections and after the hearing an Investigative Hearing Examiner (IHE) issues a recommendation to the ALRB. The IHE either recommends that the election be certified, so that the union which won a majority vote is the bargaining representative of the workers (or no union is certified and another election cannot be held for 12 months); or the IHE recommends that the election results be set aside because there was interference with employees' free choice or because of technical problems with the election. The union or the employer can appeal the IHE recommendation to the five member Board in Sacramento. The Board reviews the case and issues an opinion on the election certification.

Any party can file a charge that an ULP has been committed. The General Counsel in Sacramento investigates charges and either dismisses them or issues a complaint. Both parties are contacted by the General Counsel, and the complaint is either settled or set for hearing. An Administrative Law Judge (ALJ) presides over the hearing and issues an opinion and recommended order (remedy). The charging party (the party filing the charge) or the charged party (the party accused of committing the ULP) or both can appeal this decision to the Board. The Board reviews the case, and issues a decision. Board decisions on ULPs can be appealed to the California Court of Appeal. This court may deny a hearing if it decides the appeal has no merit. The appealing party can file a petition for review with the California Supreme Court after the Court of Appeal has either issued a decision in the case or denied a hearing on it, but the California Supreme Court is not required to hear such appeals. Under certain conditions, a federal appellate court may hear an appeal of a Board decision.

A Board certification of an election cannot be directly appealed to the courts. However, an employer can seek judicial review of an ALRB certification through indirect means, called a technical refusal to bargain. The employer contesting certification refuses to bargain with the union. The union then files an unfair labor practice charge alleging that the employer has violated ALRA Sec. 1153(e) by refusing to bargain. Since this is an unfair labor practice charge, it may be appealed to the courts. Before the Court of Appeal the employer argues that the election certification was improper and, therefore, he had no duty to bargain. The court reviews the validity of the election in the appeal. (Does a losing union have an opportunity for court review of an ALRB election decision?)

This appendix presents ALRB and court decisions on election objections and ULP cases. The Board bases its decisions on past ALRB and NLRB decisions, court opinions, and legislative intent. The Board applies the law to the facts as reported by the ALJ or IHE. The Board cannot change the findings of fact (what actually happened) determined by the ALJ or IHE unless the findings are not supported by evidence.

B. REPRESENTATION ELECTIONS

Representation elections involve high stakes for both the union and the employer. If the union wins the election and is certified by the Board, the employer must bargain with it. If the union loses, it cannot call another election on the farm for 12 months and it will have lost time and money on the campaign.

Election results may be set aside if the technical requirements for an election were not met or if workers were coerced by pre-election conduct of a party. The ALRB is more reluctant to set aside an election for improper pre-election conduct than is the NLRB. Because of the seasonal nature of agriculture, another election might not be possible for 12 months when employment again reaches 50 percent of peak.

Types of pre-election conduct which can cause the election results to be set aside include threats, interrogations, promises, and surveillance. The following case illustrates how an employer's promises can interfere with workers' free choice.

Case 1. Promises Which Influence an Election

ALBERT C. HANSEN
(1976) 2 ALRB No. 61

. . .

On September 10, 1975 a representative election was held at the Hansen Farms in which votes were cast as follows: "no union" - 224; UFW - 221; Western Conference of Teamsters (hereinafter "Teamsters") - 36; challenged ballots - 48; and void ballots - 2. Since no party received a majority of votes, a run-off election was held on September 25, 1975, the results of which were: "no union" - 300; UFW - 247; challenged ballots - 28; and void ballots - 5.

. . .

Petitioner's [UFW] principal contention is that the employer made promises of benefits to a majority of the workers in the voting unit if they voted "no union." These alleged promises were made in separate speeches by the employer to various crews while they were in the fields. It is the premise of petitioner that these promises illegally interfered with the election. Additionally, petitioner maintains that employer's agents made threats and/or promises to various workers and that these incidents in themselves are sufficient basis for setting aside the election.

The employer contends that he acted in a fair and proper manner at all times and that he was entitled under our Act to make speeches and distribute literature. He has submitted a written letter which he gave to the workers and which he contends was similar to his speeches. Prior to circulating this letter he had it reviewed by his attorney for its appropriateness under our Act. Employer Hansen testified that he never made any promises to workers with respect to wages, better working conditions or insurance. When asked what the wages would be Mr. Hansen maintains he always replied that he could make no promises because it was against the law.

. . .

At issue here are the conflicting interests of the employer's right of free speech and the employees' right to cast an uncoerced vote. The "free speech" provision of our Act is contained in Labor Code, §1155, which provides:

> The expressing of any views, arguments, or opinions, or the dissemination thereof shall not constitute or be evidence of an unfair labor practice if such expression contains no threat of reprisal or force or promise of benefit. . .

In determining whether campaign rhetoric is sufficient to set aside an election, we look not only to the nature of the speech itself but also to whether, in the light of the total circumstances, it improperly affected the result.

Upon the filing of the certification petition by the UFW on September 2, employer Hansen began a campaign for a "no union" vote. A letter dated September 3 was issued to employees urging them to vote nonunion and enumerating five "disadvantages" of a union, and extolling company policy and performance concerning wages, benefits and working conditions. This letter mentioned the ALRA and urged the workers to listen to all sides. We do not find this letter objectionable on its face. . . .

Within a three- to six-day period between the filing of the petition and the election, employer Hansen and his personnel manager, Tony Vasquez, made a set of speeches before 10 to 12 crews in the fields. . . .

Although Hansen testified that the letter dated September 3 was the basis of these speeches, testimony of workers from five different crews indicates that he deviated from the text of his carefully worded letter. He was quoted as having told the workers that if "no union" won, they would have "better wages, better benefits," "better wages or the same wages that other companies with the union would have," "best wages in the Valley," "better benefits than the union," "well, everything." Several workers testified that he said when "no union" was certified, he would negotiate a contract with representatives from each crew. Workers testified that many of these statements were prefaced with the employer's remark that he couldn't promise anything because it was against the law.

. . .

A worker from the crew of Jesus Lopez quoted Hansen as saying that in case of a UFW victory "if it was convenient for him he would negotiate, and if not there would be a strike." If "no union" won, he would pay them "higher than the other companies, and the best benefits." A member of Crew 2 quoted Mr. Hansen as saying that if "no union won he would give them "the best wages in this Valley." He said further that if Chavez won and "he [Hansen] couldn't come to an agreement with the negotiators, he wasn't going to sign . . . there would probably be a strike."

He said he could not promise them anything in writing, but if the election came out "no union," he would give the workers a list of all that he was going to guarantee them. Another worker testified that Mr. Hansen told Crew 3 "he could promise to give [them] all the benefits that any union would promise" and that "he would pay ... one cent more a carton than what the union would ask." In front of Crew 4 he was heard to say that in one year he would "give the workers more than any union." In response to a request from the workers to see a contract, Hansen replied "he would pay ... more than any union did and to please take his word for it, but that he could not show ... the contract at the time." If the workers voted "no union" he said he would "plant some more and hire more crews." In the second set of speeches, many of the employer's comments were again prefaced with the remark that he could not promise them anything.

We turn now to the objections based on threats made by Hansen supervisors. While this testimony is also in substantial conflict, we have reviewed the record as a whole and find the facts substantially as follows.

The alleged threats were made by two of the employer's supervisors, Fidel Rodriguez and Francisco Palmeno. Each of these men supervised four crews and in this capacity they were clearly agents of the employer and known to the workers as such. There was testimony that before the first election Fidel Rodriguez addressed Crew 2 saying that if the Chavez won "they would not plant any more lettuce; they would plant alfalfa ... [and] barley, because they had a lot of cattle." A member of Crew 4 testified that a week to 10 days before the first election Rodriguez told them ... that "if no union won, Mr. Hansen would plant 800 acres more of lettuce and hire two more crews." At the same time Rodriguez was also quoted as saying that if Chavez union was to win, Mr. Hansen "wasn't going to plant anything else any more; that he didn't have to, he had a lot of money anyway."

In interpreting the effect these remarks had on the employees, we note that alfalfa and barley require little if any work by farmworkers. Thus, the result of planting these crops, instead of lettuce, would be to put the lettuce crews out of work.

. . .

There was also testimony as to threatened layoffs. One worker related a conversation with Rodriguez in which he was told that people in the lettuce cutting crew would be laid off "because of the Chavez movement." This conversation was overhead by a fellow worker who then relayed it to a group of 10 to 15 other workers. The worker testified further that Rodriguez made veiled offers of a promotion to a "truckdriver" if he would "get out of that movement."

For the first time this board must consider the effect of an employer's promise of benefits to his employees made during a vigorous campaign. The National Labor Relations Board (NLRB) has characterized the issue as "the right of the employees to an untrammeled choice, and the right of the parties to wage a free and vigorous campaign with all the normal legitimate tools of electioneering." (Citation omitted.)

[The NLRB has] looked "to the economic realities of the employer-employee relationship," stating it would "set aside an election where we find that the employer's conduct has resulted in substantial interference with the election, regardless of the form in which the statement was made.". . .

[In NLRB v. Exchange Parts the NLRB] had found the announcement and granting of improved benefits prior to the election to be an unfair labor practice. Sounding the theme of an "economic realities" test, the [U.S. Supreme] Court pointed out:

> The danger inherent in well-timed increases in benefits is the suggestion of a fist inside the velvet glove. Employees are not likely to miss the inference that the source of benefits now conferred is also the source from which the future benefits must flow and which may dry up if it is not obliged. (Citation omitted.)

An "economic realities" test has also been applied to threats of reprisal and shutdown. The United States Supreme Court has articulated a distinction between legitimate economic predictions and threats of retaliations in NLRB v. Gissel Packing Co.:

> [T]he prediction must be carefully phrased on the basis of objective fact to convey an employer's belief as to demonstrably probable consequences beyond his control. . . . If there is any implication that an employer may or may not take action solely on his own initiative for reasons unrelated to economic necessities and known only to him, the statement is no longer a reasonable prediction based on available facts but a threat of retaliation based on misrepresentation and coercion, and as such without the protection of the First Amendment. (Citation omitted.) (Emphasis added.)

The purpose of any regulation of campaign conduct is to promote the free choice of employees by assisting them in making a reasoned decision as to where their own best interests lie. To this end it is important for the employee to evaluate (a) to what extent a union can improve his or her working situation and (b) what disadvantages lie in unionization. . . . Thus, the voter must be free of duress and coercion.

Our evaluation of the employer's pre-election conduct must ask first whether the conduct was an unfair use of his economic position. If the conduct is found to be objectionable, the inquiry must proceed to a determination of the effect such conduct might have had on the election. Conduct which tends to interfere with the free choice of a significant number of voters will be sufficient to set aside an election.

The objections in this case indicate a course of conduct which we have considered as a whole. (Citation omitted). We consider first the allegations concerning promises of benefits and higher wages. The employer's right to free speech would necessitate that he be allowed to recite past benefits and wage increases for which he was responsible, as well as to point out upcoming benefits and wage increases, if any, which were decided before union activity and which are not tied to the results of the election. . . .

We must view the statements of the employer in the light of the message actually conveyed. Although the substantive intent of the employer may well have been to stay within the letter of the law, it appears his speeches went further. We note that the impression received by the workers was a pattern of implied, if not actually expressed, promises contingent on the outcome of the election.

The record here indicates no basis for the campaign promises other than to influence the outcome of the election. The employer could not know what benefits or wages a union would ask for, nor could he unilaterally predict the outcome of negotiations. By making flat promises to do better for the workers than any union could do, the employer misrepresented the bargaining process and undercut the basis on which a union could campaign. Workers are especially susceptible to such statements in situations, as here, where they are deciding for the first time by secret ballot whether or not they want to be represented by a union.

The danger of benefits "which may dry up if not obliged," as pointed out in Exchange Parts, is certainly present here. The employer explicitly tied the promised benefits to the outcome of the election. After voting nonunion, however, the employees would have no means to enforce the promises which swayed their vote.

Applying the "economic realities" test of Dal-Tex Optical, we look to (1) the economic relationship between the speaker and the listener and (2) the message that was actually conveyed. The listeners in this case were in a position of economic dependence on the speaker, Mr. Hansen. The message which was conveyed to the employees was consistently one of promises of better wages and working conditions if they voted "no union." A fair evaluation of their own best interests was thus impossible in light of the coercive effect of these statements. While saying he would do better than the unions, the employer was proposing to bargain with representatives elected by the workers, thus underscoring the futility of a union vote. We find that the employer's continuous promises of benefits interfered substantially with the election, and that the election should therefore be set aside.

Additionally, we find that the alleged threats by the employer's supervisors provide another ground on which the election could be set aside. Credible and consistent testimony of employees indicated a pattern of threats of job losses if the union won the election. Whether these threats were expressed or implied is irrelevant when a clear meaning was perceived by the employees. The statements about planting alfalfa or barley instead of lettuce, the equivalent of threats of shutdowns or plant closing in the industrial setting, would be coercive conduct. Likewise, the alleged threats of layoffs in the case of a Chavez victory would have a coercive effect on the employee's vote. We are not swayed by employer's argument that no actual layoffs were made and that it was company policy to go overboard in not laying off or firing known supporters of the UFW. Coercive conduct is not limited to threats made good. In the charged atmosphere of the earliest elections under the ALRA, these threats would most certainly have an ominous effect. The threats of job losses in the case of a union victory intermingled with promises of benefits if "no union" won presented a contorted picture to employees which substantially interfered with their free choice.

* * *

Question 1a: Why might Hansen's employees be coerced by Hansen's speeches?

Question 1b: Why would a union contest an election it had lost?

Case 2. Access by Nonemployees to An Employer's Property

In order for employees to form and join unions they must be able to meet with nonemployee union representatives and discuss union issues. In the context of seasonal work, migrant workers and seven-day elections, the only feasible place for union organizers to find the workers and meet with them is the workplace. The ALRB, therefore, has promulgated a set of regulations which allow limited access to an employer's property by union organizers. Access is limited by time (one hour before work begins, one hour at lunch, and one hour after work ends—for four periods of 30 days in any one year) and by number of organizers (two organizers for the first 30 members of the crew and one additional organizer for each 15 additional crew members).

Some growers asserted that this right of access violated their property rights. The California Supreme Court held that the access regulation was a constitutional use of the ALRB's power and the U.S. Supreme Court dismissed the appeal of this case. The opinion in ALRB v. Superior Court lays out the NLRA precedent which led to the ALRB's conclusion that access was a necessary balancing of rights in the agricultural context in order for employees to enjoy fully the right of self-organization.

AGRICULTURAL LABOR RELATIONS BOARD V. SUPERIOR COURT
(1976) 16 C. 3d 392

. . .

The issue joined here is new to the California courts, but our federal brethren have often considered it in the industrial labor context. . . . "In Republic Aviation Corp. v. Board (citation omitted) the Supreme Court set forth the ground rules concerning union activity on company property." (Citation omitted.) The case dealt with organizational activities conducted on the employer's premises by union spokesmen who were also employees of the company. The high court ratified the position of the NLRB that absent extraordinary circumstances it is an unfair labor practice for the employer to prohibit such activities during nonworking hours. . . . "It is not every interference with property rights that is within the Fifth Amendment. Inconvenience, or even some dislocation of property rights, may be necessary in order to safeguard the right to collective bargaining." (Citation omitted.)

The second landmark case on this topic is Labor Board v. Babcock & Wilcox Co. (Citation omitted.) In contrast to Republic Aviation, the union organizers excluded from the employers' premises in the three consolidated cases decided in Babcock & Wilcox were not employees of the companies in question. . . . The Supreme Court ruled that the board erred in failing to draw a distinction between employee and nonemployee organizers: access to company property by the latter can be denied, said the court, "if reasonable efforts by the union through other available channels of communication will enable it to reach the employees with its message." (Citation omitted.)

By declaring the foregoing standard the court message rejected any claim that "property rights" of employers are paramount to their employees' right to have effective access to information assisting them in their goal of self-organization: "The right of self-organization depends in some measure on the ability of employees to learn the advantages of self-organization from others." (Citation omitted.) Rather, employers' property rights must give way whenever the two interests are found to be in irreconcilable conflict. . . . "[W]hen the inaccessibility of employees makes ineffective the reasonable attempts by nonemployees to communicate with them through the usual channels, the right to exclude from property has been required to yield to the extent needed to permit communication of information on the right to organize." (Citations omitted.)

Examples of the application of this rule appear in a variety of contexts. In Republic Aviation the court in dictum distinguished the case before it from those involving a mining or lumber camp where the employees pass their rest as well as their work time on the employer's premises, so that union organization must proceed upon the employer's premises or be seriously handicapped. (Citations omitted.)

Shortly thereafter such a case arose.   In National Labor Rel. Bd. v. Lake Superior Lumber Corp. (citation omitted) the employer operated a number of lumbering camps on its timber tract. Each was isolated from any town, and was largely self-sufficient. The employees lived on the camp premises in bunkhouses; although given Sundays off, they usually remained in the camps. In these circumstances the NLRB ruled it was an unfair labor practice for the employer to bar nonemployee union organizers from entering the bunkhouses to talk with the men during nonworking hours. . . .

Nor is the right of access limited to remote lumber or mining camps; it may attach in the case of a ship anchored in a busy port.   Thus, in National Labor R. Board v. Cities Service Oil Co. (citation omitted) the employer operated ocean-going oil tankers which entered United States ports to discharge their cargo. A maritime union was refused passes to board the ships while in port for the purpose of negotiating grievances of the seamen. The NLRB ruled this practice violated the seamen's rights to self-organization and collective bargaining under section 7 of the NLRA. The Second Circuit Court of Appeals agreed, reasoning that "The result of refusing passes is undoubtedly to prevent the most effective sort of collective action by the employees. Ships, and particularly these oil tankers, which ordinarily remain in port for a day only, afford less opportunity for investigation of labor conditions than do factories where the employees go home every afternoon and have the evenings at their disposal. There is no cessation of work at the end of each day for seamen on a tanker. A large number of them are on watch, others are loading or discharging cargo; their hours for work and shore leave are different and, in the short time the vessel is in port, it is impossible for Union representatives to assemble the unlicensed personnel either on shore or on shipboard to discuss grievances or investigate conditions. The Union must have the members of the crew readily accessible in order to work to any real advantage. . . ." The court therefore granted enforcement of the board's order of access. (Citations omitted).

The same result has been reached on a showing of significantly less employee isolation than in the foregoing cases. In NLRB v. S&H Grossinger's Inc. (citation omitted) the employer operated a large rural resort hotel located only one and one-half miles from the nearest town. Sixty percent of the employees lived on the premises, but the remainder lived in neighboring towns and drove to work by car or taxi.   The employer refused access to its premises by nonemployee union representatives, and the NLRB ruled this to be interference with the employees' right of self-organization. The federal circuit court observed that "No effective alternatives are available to the Union in its organizational efforts. The resident employees have no telephones in their rooms. Radio and newspaper advertising are expensive and relatively ineffectual. Moreover as far as radio is concerned, there was no single time at which a major proportion of employees would be off duty and free to listen to a message broadcast by the Union. ..."

The court then quoted and applied the principles of Babcock & Wilcox as follows: "Here the majority of the employees live on the employer's premises. They cannot be reached by any means practically available to union organizers. As against these considerations Grossinger's raises only its proprietary interest. It shows no detriment that would result from the admission to its property of the Union's representatives under those reasonable regulations as to place, time and number which the Board's order contemplates.

...

Thus the rule of Babcock & Wilcox, both as enunciated and as applied, is clear: if the circumstances of employment "place the employees beyond the reach of reasonable union efforts to communicate with them, the employer must allow the union to approach his employees on his property." (Italics added.) (Citation omitted.) This language could not be plainer. We deem it dispositive of the issue of the federal constitutionality of access to agricultural property under the challenged regulation of the ALRB (citation omitted) and of the claim of invalidity premised on the cited provisions of the California Constitution. (Art I §1,7, subd. (a) and §19). In the present context we construe those sections to guarantee no greater rights to California property owners than do their federal counterparts.

The only remaining question in this regard is whether it is constitutionally required that the determination of employee inaccessibility within the meaning of the Babcock & Wilcox test be made on a case-by-case basis, as the real parties urge, rather than by a rule of general application. As will appear, there is not authority for imposing such a requirement as a matter of constitutional law.

[After a discussion of several constitutional principles, the Court states:] We conclude that the decision of the ALRB to regulate the question of access by a rule of general application transgresses no constitutional command.

...

... [The] ALRB predicated its access regulation on factual findings phrased on the very language of Babcock & Wilcox. Those findings disclose that the board did not adopt the NLRB practice on the access question because it determined that significant differences existed between the working conditions of industry in general and those of California agriculture. As we have seen, in regulating industrial labor disputes the NLRB has authorized access by union organizers to employers' premises when, for example, the same employees did not arrive and depart every day on fixed schedules, there were no adjacent public areas where the employees congregated or through which they regularly passed, and the employees could not effectively be reached at permanent addresses or telephone numbers in the nearby community, or by media advertising.

By contrast, the ALRB found that such conditions are the rule rather than the exception in California agriculture. The evidence heard by the board showed that many farmworkers are migrants; they arrive in town in time for the local harvest, live in motels, labor camps, or with friends or relatives, then move on when the crop is in. Obviously home visits, mailings, or telephone calls are impossible in such circumstances. According to the record, even those farmworkers who are relatively sedentary often live in widely spread settlements, thus making personal contact at home impractical because it is both time-consuming and expensive.

Nor is pamphleting or personal contact on public property adjacent to the employer's premises a reasonable alternative in the present context, on several grounds. To begin with, many ranches have no such public areas at all: the witnesses explained that the cultivated fields begin at the property line, and across that line is either an open highway or the fields of another grower. Secondly, the typical industrial scene of a steady stream of workers walking through the factory gates to and from the company parking lot or nearby public transportation rarely if ever occurs in a rural setting. Instead, the evidence showed that labor contractors frequently transport farmworkers by private bus from camp to field or from ranch to ranch, driving directly onto the premises before unloading; in such circumstances, pamphleting or personal contact is again impossible. Thirdly, the testimony established that a significant number of farmworkers read and understand only Spanish, Filipino, or other languages from India or the Middle East. It is evident that efforts to communicate with such persons by advertising or broadcasting in the local media are futile. Finally it was also shown that many farmworkers are illiterate, unable to read even in one of the foregoing languages; in such circumstances, of course, printed messages in handbills, mailings, or local newspapers are equally incomprehensible. . . .

In addition, the problem here is compounded by the provisions of the ALRA which require swift elections—a difficulty not faced by the NLRB. In all cases involving crops with short harvest seasons, the union petitioning for the election has only a brief time in which to gather the necessary employee signatures (Lab. Code §1156.3, subd. (a).) An intervening union will have even less time—at most 6 days—to obtain the signatures of 20 percent of the workers in order to qualify for the ballot. (Id., subd. (b).) And both unions have only a few days thereafter to explain their positions to the workers. In such circumstances most of the channels of communication which have been used in organizing industrial laborers, and which were found sufficient in <u>Babcock & Wilcox</u> and its progeny, are simply too slow to be effective. . . .

. . .

We conclude from the foregoing that the decision of the board to create a limited right of access by means of a detailed and specific regulation does not conflict with any intent of the Legislature inferable from its enactment of sections 1148 and 1152.

* * *

Question 2. What is the basic difference between the NLRB access rule, as set forth in the Babcock & Wilcox case and the ALRB access rule? What are the reasons for the difference?

C. DECERTIFICATION ELECTIONS

Workers can decertify their collective bargaining representative only by filing a decertification petition and then obtaining a majority vote for "no union" in the decertification election. Since the election determines employee desires, an employer cannot instigate or assist a decertification campaign. Employer assistance can be grounds for ALRB refusal to certify a decertification election. Employer assistance in decertification elections has generated large makewhole awards if the employer stops bargaining with the union because it believes it "won" a decertification vote. A major issue in many decertification elections is the exact definition of employer assistance.

Case 3. Employer Assistance

ABATTI FARMS, INC. AND ABATTI PRODUCE, INC.
(1981) 7 ALRB No. 36

. . .

Respondent (Abatti) first excepts to the ALO's conclusions that respondent instigated and assisted the employees' decertification efforts, in violation of §1153(a) of the Act. Since instigation and unlawful assistance are separate concepts (citations omitted.), we shall analyze the evidence as to each separately. While we conclude that the record does not warrant the ALO's conclusion as to instigation, it amply supports his conclusion as to unlawful assistance.

The ALO inferred the fact of instigation from his consideration of a number of factors, including doubts about whether the Petitioners were sufficiently motivated to undertake such a campaign. Both men, however, resisted joining the union pursuant to the contract's union security clause and although Cruz finally did join the union, he did so only after he was threatened with the loss of his job; Castellanos never did join and was discharged. While there is some doubt about the veracity of Castellanos' version of how he came to know about decertification procedures, .. . in view of petitioners' trial-and-error approach in getting the petition underway and of the disaffection of Respondent's regular employees for the union, the evidence falls short of establishing that Respondent implanted the idea of decertification in the minds of Petitioners. (Citations omitted.)

There is, however, ample evidence of Respondent's unlawful assistance to the employees in their decertification efforts. In considering such evidence, we first note that "... in a case of this kind, involving a charge of violations of the duty not to maintain a forbidden relation, a reliance on so-called circumstantial evidence is not only permissible, but often essential. On the very nature of such a case, there will seldom be discoverable data showing direct statements by a party charged with violation that he has performed improper acts." (Citations omitted.) In this case, the circumstantial evidence is of two kinds: the first goes to prove that the leading proponents of the decertification petition were provided leaves of absence and other benefits to facilitate their conduct, or as a result of their conduct, of the campaign; the second goes to prove that Respondent's agents assembled its employees for the purpose of obtaining signatures on the various decertification petitions. . . .

The record supports the ALO's finding that Cruz was personally assisted by Respondent. . . . His extended absence from work to circulate the petitions, his receipt of a Christmas bonus well in excess of any bonus received by the other tractor drivers, Respondent's allowing him to charge Respondent for the broken glass on his car and waiting to deduct the cost from his paycheck until shortly before the hearing, and his eligibility for insurance even though he did not work enough hours during the month he was circulating the petition to entitle him to coverage are all factors which support the conclusion that Respondent not only permitted Cruz to campaign, but also abetted him in his decertification efforts by insuring that he lost nothing because of the time he spent campaigning.

Respondent attempts to dispel the inference of unlawful interference readily drawn from these facts by pointing to evidence of its "liberal" leave policy, its history of largess to employees in emergencies and its practice of permitting employees to charge personal items. Respondent's ordinary practices differ sufficiently from its treatment of Cruz in these respects to compel the conclusion that Cruz received special favorable treatment because of his involvement in the decertification campaign. Thus, Respondent's treatment of Rosa Briseno's leave for union business stands in stark contrast to its assertations of a "liberal" leave policy with respect to an employee's concerted activities. Its failure to deduct from

Cruz' pay the cost of repair to his auto immediately after it was incurred is also atypical, as is the size of the Christmas bonus it gave Cruz.

Finally, there is the matter of Respondent's making the arrangements which resulted in Petitioners' being represented by counsel. Although the Act cannot require an employer to refuse to respond to employee inquiry (citations omitted.), the evidence in this case shows that Respondent went well beyond merely naming or suggesting a lawyer whom Petitioners might consult; it brought Petitioners and counsel together. . . .

We also agree with the ALO that Jose Rios unlawfully assisted in circulating the decertification petition and that Respondent's giving a Christmas party, at which time the decertification petition was circulated in the presence of its supervisors, also constituted unlawful support of the decertification effort. With respect to the campaigning in the fields, the evidence is, as the ALO notes, sharply conflicting. The ALO resolved these conflicts in conformity with his credibility resolutions and based on our review of the record we defer to his conclusions.

In light of our finding of unlawful assistance, we do not face the question whether an employer may rely on good faith doubt of majority status when a decertification petition raises a real question concerning representation. The general rule is that there is no good faith in a doubt which an employer has manufactured:

> [Respondent] cannot, as justification for its refusal to bargain with the union, set up the defection of union members which it had induced by unfair labor practices, even though the result was that the union no longer had the support of the majority. It cannot thus, by its own action, disestablish the union as the bargaining representative of the employees, previously designated as such by their own free will. (Citation omitted.)

The rule applies with equal force to decertification campaigns: an employer that has orchestrated "a union-busting campaign cannot rely on the pendency of a decertification petition or the loss of majority status to justify [its] refusal to bargain." (Citation omitted.)

. . .

Dismissal of Petition for Decertification

Because of Respondent's support and assistance of the decertification campaign, the Petition for Decertification shall be, and it hereby is, dismissed.

* * *

Question 3a. The vote tally in the decertification election at Abatti Farms was:

$$\begin{aligned}
&\text{UFW} \dots\dots\dots\dots\dots 125 \text{ votes}\\
&\text{No Union.} \dots\dots\dots\dots .149 \text{ votes}\\
&\text{Challenged Ballots.} \dots\dots .121 \text{ votes}
\end{aligned}$$

Give three reasons why this decertification vote should not be certified.

Question 3b. If the vote were lopsided, say No Union-300 and UFW-100, would the Board decision have been different?

D. SUCCESSORSHIP

When a union is certified as the collective bargaining representative of an employer's farmworkers, the employer has a duty to bargain with the union. If the employer later sells its farm, an issue arises over whether the new employer is a successor to the certified employer and therefore has a continuing duty to bargain with the union. The ALRB and the Supreme Court agree that there is no easy test to determine whether or not an employer is a successor; the answer depends on balancing the rights of the employer against the rights of the workers. The ALRB announced the successorship factors it considers in Highland Ranch/ San Clemente Ranch.

Case 4. Successorship

HIGHLAND RANCH/SAN CLEMENTE RANCH
(1979) 5 ALRB No. 54

. . .

. . . In the instant case, as no contract existed between Highland and the UFW as of the date the business was sold, we are concerned only with whether the successor employer, San Clemente, had a duty to bargain with the UFW as the chosen representative of the employees.

Given the unusual characteristics of agricultural ownership patterns and the agricultural labor force, . . . as described above, an approach to successorship which examines factors in addition to the continuity of the work force is most appropriate. Undue emphasis on the continuity of the work force factor at the expense of other relevant factors would render the important protection provided employees by the successorship principle almost entirely ineffective. We will, therefore, not ignore this factor but will give careful consideration to other factors as well.

Highland completed the sale of its leasehold interest . . . and equipment to San Clemente on November 29. San Clemente took possession of the land and equipment on December 1. This date coincided with the end of the tomato harvest. On December 4, San Clemente hired four individuals, three supervisors and one irrigator, who had been employed by Highland. On December 9, the UFW requested bargaining with San Clemente, which rejected the request. By the end of February 1979, San Clemente had a work force of 49, of which 46 had been Highland employees. On March 14, 1978, it began obtaining some of its employees through a labor contractor. By March 25, 1978, it had 150 employees, 42 of whom were supplied by the labor contractor and 70 of whom were former Highland employees. San Clemente continued to grow basically the same crops on the same land and processed them at the same packing shed.

The fluctuating size of the work force at this operation is typical of California agriculture. In the face of such fluctuations, we are unwilling to adopt the position urged by San Clemente that, first, whether it is to be regarded as Highland's successor for purposes of the Act should depend solely on the number of former Highland employees who were in San Clemente's work force when it had hired its "full complement" of employees and, second, that the "full complement" of employees means 50 percent of the workers it employs at its peak employment period. A rigid, mechanical rule of this sort is ill-suited to the complex realities of the industry and would be a poor substitute for the exercise of judgment on the particular facts of each case.

In this case, because the sale occurred in the "off-season," the only employee that was apparently needed on December 1, when San Clemente took possession, was an irrigator to maintain cabbage that had been planted, the tomato harvest having just ended. As the crop grew and needed additional care, and as other crops were planted, the work force continued to increase until harvest and the ensuing layoffs. Unlike the industrial setting, an agricultural employer's full complement of employees can vary from day to day and season to season. On December 1, the Employer's full complement of workers may well have been only four or even one. To permit a succeeding employer to abolish the rights of his predecessor's employees by the hiring of one or two individuals would make a mockery of the principle that employees' collective bargaining rights are entitled to protection when the ownership or structure of an enterprise is changed. . . . On the other hand, the time elapsing between an off-season sale and the first subsequent peak may generally be considerable. An employer who is subsequently found to be a successor would thus be in a position to evade its duty to bargain for a significant period of time during which collective bargaining should already have been underway. Again, this dilemma is brought about by the seasonality and high turnover prevalent in the agricultural context.

Despite the transfer of ownership from Highland to San Clemente, the agricultural operation itself remained almost identical. There was no significant alteration in the nature of the bargaining unit. (Citation omitted.) Unit employees perform the same tasks for San Clemente which they previously performed for Highland since San Clemente grows essentially the same crops. The size of the unit also remained the same. Furthermore, San Clemente is farming the same land as Highland, having acquired the lease to all of Highland's agricultural property. It has also acquired Highland's agricultural machinery which it uses in its farming operations. In these circumstances, meaningful principles of successorship can be given effect only by finding that San Clemente is Highland's successor. For us to reach the contrary result would be to miss the forest for the trees. Accordingly, based upon all of the above functions, we find that San Clemente is a successor to Highland and that it violated its duty to bargain with the UFW by refusing to meet and to supply relevant information. . . .

* * *

Question 4. List five factors which the Board considered in deciding that San Clemente is a successor to Highland.

E. DISCRIMINATORY CONDUCT

Sector 1153(c) of the ALRA prohibits an employer from "discrimination in regard to the hiring or tenure of employment, or any condition of employment, to encourage or discourage membership in any labor organization."[1] The four elements of proof required in an 1153(c) case are (1) a discriminatory act; (2) union activities by the discriminatee; (3) employer knowledge of the discriminatee's union activities; and (4) anti-union motivation.

The discriminatory act is often the firing or refusal to rehire a worker. However, discrimination in any condition of employment is prohibited by §1153(c), so violations have included discriminatory demotion, assignment to less desirable work, eviction from company housing, and withdrawing of privileges. The following case involves a discriminatory policy of denying a crew the opportunity to purchase harvest-related equipment at a discount. The company was held responsible for the unfair labor practice of the farm labor contractor (FLC), since FLCs are not considered "employers" under the ALRA.

Case 5. Discrimination Because of Union Activities

BAKER BROTHERS/SUNKIST PACKING HOUSE
(1986) 12 ALRB No. 17

. . .

Respondent's Operation

Baker Brothers is a partnership engaged in citrus production in Tulare County. It not only grows and harvests its own citrus, but it provides harvesting services for other growers and operates a packing shed where its citrus and that of other growers is packed and shipped. The partnership has been in existence since 1960. The partners are four bothers; one brother, Leland "Hap" Baker, is the General Manager and another, Tom Baker, is the Field Supervisor.

The harvesting service which respondent operates for itself and most of the growers for whom it packs utilizes the services of a labor contractor. For the past ten years that contractor has been Conrad Sanchez. In recent years, Sanchez and his partner, Joe Diaz, have provided Baker Brothers with two harvest crews. Domingo Barba is the foreman of one and Nickolas Balderas is the foreman of the other. Barba has worked for Sanchez as a crew foreman since 1965 and has been harvesting at Baker Brothers for the past 15 years. Balderas has been a crew foreman for Sanchez since 1976 and he has been harvesting at Baker Brothers since June 29, 1981.

The two crews primarily harvest Navel and Valencia oranges. The Navel season runs from November through March, and the Valencia season runs from April through October. In 1981 the Barba crew worked 188 days for Baker Brothers while the Balderas crew worked only 30 days. In 1982 the Barba crew worked 152 days while Balderas' crew worked only 35 days. However, in 1983 (January 1 to October 29) the Barba crew worked 124 days and the Balderas crew worked 137 days. . . . During most of the three year period there were 30 to 35 workers in each crew. Because oranges need not be harvested as soon as they ripen, crew utilization is more dependent on the condition of the market than the maturity of the crop. However, the operation of normal market forces is circumscribed by Marketing Orders which allocate to each handler a pro rata share of Navel and Valencia production for domestic consumption which may not be exceeded.

The groves in which the crews work are scattered about the packing shed at an average distance of 10 miles. The determination of which groves are to be harvested is made by the shed each evening for the following day. Normally, Tom Baker contacts Conrad Sanchez, and Sanchez, in turn, calls his foremen who then notify their crews of locations and starting times. Work begins early in the morning during the summer so as to avoid the afternoon heat and late in the morning during winter so as to avoid the damp morning fog. Tom Baker also arranges for bins to be trucked to and distributed within the groves so that work can begin promptly and so that each crew will have enough bins on hand to handle its daily pick. Workers are paid a piece rate based on the number of bins they fill each day.

. . .

The Union Organizational Drive and Election

A union certification election was conducted among the employees of the respondent and its labor contractor on March 11, 1983. Fifty-nine names appeared on the eligibility list; 51 workers voted for the UFW, and eight voted for no union. Fourteen workers voted subject to challenge (13 of them because their names were not on the list). . . .

The dissatisfaction which led to the UFW's organizational drive and its pronounced success in the election began in Domingo Barba's crew. Crew members were unhappy with the uncertainty of the piece rate, the inadequacy of toilet and clean-up facilities, the demand that bins be filled fuller, and the delays experienced in being informed of daily locations and starting times. As a result, crew member Jose Zacarias contacted Barbara Considine of the UFW, and he and his wife, Carmen, met with her in their home to discuss union representation. Considine went on to hold three or four meetings at the Zacarias' home with members of both of the contractor crews. A majority of the Barba crew and a third of the Balderas crew attended. A negotiating/grievance committee made up of three members of the Barba crew and two members of the Balderas crew was established. On March 4, 1983, Considine filed a Notice of Intent to Organize, a Notice to take Access, and a Petition for Certification. She then began to visit the crews in the groves where they worked. On her first visit Barba questioned her right to be there, but relented under pressure from the crew. During this and in later visits she spoke with workers and handed out UFW authorization cards and buttons. A majority of the Barba crew began wearing the buttons. She also led the crew in a "union clap"— clapping slowly and continuously in unison at an increasing rate. Barba was the only supervisor present during Considine's visits.

Baker Brothers retained the Farm Employers Labor Service to help persuade workers to vote against the UFW. Two of the Service's labor consultants, accompanied by Joe Diaz, visited the Barba crew and explained that their employer was concerned for their welfare and wanted to do well by them. When the consultants told the workers that they already had a good medical plan, the Zacarias' challenged them, saying that they were unaware of any medical coverage. The consultants said they would be back later that day with specific information. A few days later, they returned with identification cards for some crew members. They also produced a "Explanation of Benefits" form indicating that the plan had paid for chiropractic services rendered to Jose Zacarias in 1978. Zacarias told them that he did not recall the claim.

In the pay envelope distributed to each worker just before the election was a sheet with a drawing of a vulture—a parody of the UFW eagle—and a statement that the UFW would take 2 percent of each employee's earnings if it were successful.

A number of members of the Barba crew attended the pre-election conference, and a lesser, but still significant number attended from the Balderas crew.... Hap Baker attended for the respondent.

Further Union and Concerted Activity: Involvement in ALRB Processes

In May 1983, a few months after the election, a dozen or so members of the Barba crew went to Sanchez' office to complain about discrepancies between the hours they actually worked and those appearing on their paychecks as having been worked for the purpose of qualifying for medical coverage. Carmen Zacarias acted as spokesperson in relaying the crew's concern to Sanchez' bookkeeper, Audrey Lugo, who was responsible for preparing the payroll and calculating hours worked. About the same time, Carmen also complained to Tom Baker that the crew frequently had to wait to begin work because bins arrived late at the groves.

Early in 1984 members of the Barba crew started once again wearing union buttons and hats on the job; and, in February, the crew staged a demonstration of their support for the union by placing UFW flags on their cars. Hap Barker was present at the time, and Sanchez drove by while the demonstration was in progress.

Late in February, Aureliano Rodriguez filed a charge with the ALRB Regional Office in Delano alleging that the respondent was discriminating against the crew by "dividing the work in a fashion that favored another crew which is made up of non-union supporters." The charge was eventually dismissed for failure of proof. . . .

In October 1984, a group from the Barba crew twice visited the packing shed to express their concern over not being recalled from layoff. During their second visit they spoke with Hap Baker. The crew was recalled a few days later. In December 1984, about ten crew members attended the ALRB hearing on the election objections, but none testified.[2]

The Respondent's Attitude Toward Unionization

Hap Baker freely acknowledged his awareness of the pro-union sympathies of the workers, but when questioned about which crew was the more active, answered:

I don't know one of those guys from another. I don't know who's the most—they both voted about the same way .[t]hey were both union crews.
. . .

He appears to have told the crew members who met with him in October 1984, that the shed had lost some growers as a result of the union drive, and he acknowledged that, "[A]lot of growers [for whom Baker Brothers harvested, packed and shipped] didn't like the union. . . . [T]hat's no secret." . . . As noted earlier, Baker Brothers hired a firm of labor consultants in the hope of defeating the union.

Barba likewise acknowledged that his crew was very pro-union at the time of the election. . . .

More revealing of the respondent's attitude toward unionization was the increase it made in the piece rate the week before the election. Up until then, it had been paying a variable rate ranging from $9 per bin in good groves (ones with a lot of big oranges) to $11 per bin in bad groves (one with few, small oranges). . . . A week before the election this was increased to a constant $10 per bin. . . . An examination of the payroll records for the two months preceding the change discloses that the $10 rate was indeed an increase and not simply an averaging of the variable rates. . . . Hap Baker testified that he knew of no particular reason why the increase was given, other than that the $10 rate had become common throughout the area.. . .

An increase in wages during a union organizational campaign carries with it a strong inference that the increase came in response to the campaign and was not, as Baker claimed, simply a way of staying abreast of area standards. (Citations omitted.) To overcome such an inference, a respondent must demonstrate that the increase comported with its customary practice of making periodic wage adjustments, or that it came as a result of a real business necessity. (Citations omitted.) Hap Baker's explanation falls short of either justification, and so the increase can be considered as an indication of respondent's background animus towards unionization. . . .

The Failure to Provide Equipment

Findings of Fact

For some time Sanchez had maintained a store of gloves, clippers, protective sleeves and picking sacks which crew members could purchase for cash or, more often, by payroll deduction. . . . Usually, they would let their foreman know what they needed; then, on his next trip to the office, he would obtain it from the bookkeeper, Audrey Lugo, and provide it to requesting workers, noting the amounts to be deducted from their paychecks on his time records. The procedure was convenient for the workers, and the prices were less than they would pay elsewhere.[3]

Immediately after the election, members of the Barba crew were no longer able to obtain equipment in this manner . . . instead, they had to purchase it elsewhere on their own time and at additional expense. . . .

The General Counsel maintains that the existing arrangement was terminated in swift retaliation for the crew's union sympathy and vote. The respondent acknowledges that crew members no longer received equipment, but attributes it to an unfortunate misunderstanding between Barba and Lugo which supposedly occurred shortly after the election when he asked her for gloves and was told that there were no more, meaning that she was temporarily out; he misunderstood and thought she meant that gloves and other equipment would no longer be provided.

Analysis and Further Findings of Fact

Respondent's "unfortunate misunderstanding" explanation looks neat enough, but, for a number of reasons, does not wear well.

First of all, there are real problems with Barba's testimony. When called by the General Counsel as an adverse witness, he began by admitting that Sanchez had told him to stop issuing equipment. . . . He was then asked whether he had admitted to workers that it was done to punish the crew. He denied ever saying that, but added, "We all thought that [was the reason]. . . ." Then, in his first, hasty about-face, he denied believing that the crew's pro-union sympathies had anything to do with it. . . . Shortly afterwards, he again reversed earlier testimony and denied that Sanchez had told him that there would be no more equipment. . . . He went on to say that even though the crew repeatedly pressed him, he had not pursued the matter with Sanchez and had never sought an explanation. . . . Under examination by respondent's counsel, he went further still and said that he had never spoken with Sanchez, but had relied entirely upon Lugo's statement that, "There was no longer any more." . . . When asked whether he understood her to mean "for that particular time or forever more," he answered:

> She told me that there was no more. That's it. I didn't know if that was for all or what. . . .

But then he testified:

> Well, she no longer told me that there was any, so I figured it was forever.
> . . .

After studying Barba's testimony in the transcript and having observed his demeanor while giving it, I am impelled to conclude that he was vainly striving to present a "line" which he did not fully comprehend, and so could only stumble through. I believe neither him nor it, and I accept Carmen Zacarias' testimony that he admitted to her that Sanchez had told him that equipment would no longer be made available to the crew. . . .

Respondent's "unfortunate misunderstanding" explanation also leaves unexplained a similar change which occurred later on in the other crew. The payroll records—the best indicator of what happened and when it happened—show that Balderas' crew continued to obtain pay advances against equipment purchases up until the week ending June 4, 1983, at which point they ceased. . . . Neither Balderas, nor Lugo, nor Sanchez offered any coherent explanation for the sudden halt. The best Balderas could come up with was: "Well, they just ask for it very seldom.". . . Lugo said only: "They haven't been in for quite awhile. I'm not sure how long its been." . . .

Given the convenience and financial savings of purchasing through Sanchez, I cannot believe that the crew simply lost interest in the arrangement. The more persuasive explanation is the General Counsel's; namely, that Balderas' equipment was cut off as a result of the crew's concerted refusal to pick a grove in April or May. The parallel between that and sudden cut off of equipment after the election to the more active Barba crew is obvious.

Furthermore, a witness false in one part of his or her testimony is to be distrusted in others. (Citation omitted.) Because I do not accept Lugo's testimony as to what happened with Balderas, I discredit her related testimony about Barba, and the same is true of Sanchez. . . .

Hap Baker denied instructing anyone to refuse to provide equipment to the crews, and I have no reason do doubt him. The equipment arrangement was between contractor and crew. There was no reason for Baker to be involved or informed. Baker Brothers' fault, therefore, is confined to the vicarious responsibility which, under the provisions of our Act, is ascribed to employers for the conduct of their contractors. (Labor code section 1140.4(c).)

Conclusions of Law

The convenience and savings available to crew members by purchasing their equipment from Sanchez is a significant term or condition of employment, such that its discriminatory elimination would constitute a violation of the Act. . . . For the elimination to be discriminatory, the General Counsel must establish that: (1) the crew had been involved in union activity; (2) that respondent was aware of its involvement; and (3) that there was a causal connection between the elimination of equipment and the union activity of the crew. (Citations omitted.)

Here, there is no question that the crew had supported the UFW during the election campaign or that the employer was aware of its support. Respondent concedes this, but argues that both crews were active, so that treating one differently than the other would be inconsistent with an anti-union motive. However, there is evidence that Barba's was somewhat more active and that Sanchez, if not Baker, knew it. Then, too, Balderas' crew did not go unscathed; eventually, it too was deprived of equipment, and for an equally illegitimate reason. . . .

As for the causal connection between the crew's union activity and the elimination of equipment, there are a number of factors which go to establish it: First is the timing, coming as it did on the heels of the election. Second is respondent's anti-union stance during the campaign, particularly its increase in the piece rate. Third is the absence of any justification which can stand up to careful scrutiny; respondent's "unfortunate misunderstanding" explanation is, for reasons already explained, unacceptable.

I therefore conclude that respondent violated section 1153(c), and derivatively section 1153(a), by terminating the practice of offering equipment for sale to the members of the Barba crew. . . .

* * *

Question 5. What is the evidence in Baker Brothers to support the four elements of proof for 1153(c) violation?

Case 6. Dual Motivation

When more than one motive is advanced for discriminatory conduct (dual motivation cases), the ALRB has adopted the test set out by the NLRB in Wright Line, Inc. 251 NLRB 150 (1980) and by the California Supreme Court in Martori Bros. Distributors v. ALRB. The General Counsel has the burden of showing that protected activity was a motivating factor in the employer's decision. The burden then shifts to the employer to show that the same decision would have been made even absent the protected activity.

The following excerpt illustrates an application of the Wright Line rule in a refusal to rehire case.

NISHI GREENHOUSE
(1981) 7 ALRB No. 18

. . .

The Significance of the Wright Line Case

Wright Line, Inc., supra, 251 NLRB No. 150 was decided by the NLRB on August 27, 1980. It was a direct effort by the NLRB to establish a clear standard for placing the burden of proof and determining causality in cases alleging violations of section 8(a)(3) of the National Labor Relations Act (NLRA). . . .

. . .

The major causality tests used by the NLRB and the federal courts prior to Wright Line were variously characterized as the "in part" test (whether a discriminatory motive was a basis for the employer's adverse action); the "but for" test (whether the adverse action would have been undertaken "but for" the employee's protected activity); and the "dominant motive" test (whether the discriminatory motive was the "dominant motive" for the adverse action) . . . Wright Line, in effect, combines certain elements of these tests by the following formula: if the General Counsel establishes that protected activity was a motivating factor in the employer's decision, the burden then shifts to the employer to prove that it would have reached the same decision absent the protected activity. . . .

. . .

Application of the Wright Line Standard to the Instant Case

Respondent's business operation is a small carnation nursery in Salinas with approximately eight employees, and various relatives of Mr. Nishihakamada (referred to by all parties as Mr. Nishi), the owner, working in close proximity in four greenhouses. The two discriminatees, Luis Batres and Jose Bernal, had worked for Respondent since 1973 and 1978, respectively. In the summer of 1979, the UFW began an organizing drive among the employees of various nurseries in the Salinas area, including Nishi Greenhouse. Batres and Bernal were the principal contacts between the union and the other employees of Respondent; they were involved openly in discussions about unionization among the other workers and in the distribution of union leaflets. The ALO found that the two discriminatees were involved in union activities and that Respondent was aware of those activities.

The parties stipulated that in January 1980, both discriminatees were charged by the Immigration and Naturalization Service (INS) with violation of the immigration laws. When they were released by the INS and returned to Respondent's premises a week later, they were refused rehire on the basis of a company policy against hiring people who had been picked up by INS.

The ALO found that Respondent engaged in certain conduct, between the initial union activity of Batres and Bernal and Respondent's subsequent refusal to rehire them, which indicated Respondent's concern about that union activity. . . . The anti-union tactics employed by Nishi included meetings held to solicit employee grievances or complaints about working conditions, a wage increase in September 1979 of 40 cents per hour (compared to past increases of five to fifteen cents per hour), and the institution of a new medical insurance plan in November 1979. After Batres served the UFW's Notice of Intent to Take Access on Respondent, Nishi instituted a "no-talking" rule and separated the workers in the different greenhouses to minimize such communications.

The solicitation of grievances, granting of unusual benefits, and changing company rules to prevent communication about union matters, all during a pre-election campaign, have been found to constitute unlawful interference with employee rights under section 1152 and are therefore evidence of anti-union animus. (Citations omitted.) We find that Respondent here demonstrated significant anti-union animus.

Based on her findings that the discriminatees were engaged in union activity, that Respondent knew of their activity and was concerned about this activity (having taken serious steps to curtail such activity), the ALO concluded that the employees' union activity was a "significant motivation" in Respondent's decision not to rehire Batres and Bernal. We find support for these findings and conclude that the General

Counsel has shown that protected activity was a motivating factor in Respondent's decision to refuse rehire to Batres and Bernal.

Respondent's Defense

Following the Wright Line test, the burden now shifts to Respondent to show that Batres and Bernal would not have been rehired, even in the absence of their protected activities.

Respondent contends that its refusal to rehire was justified by a business policy of not rehiring workers who had been picked up by the INS. Nishi testified that this policy was established in August or September 1979, following a meeting of nursery owners in the Salinas area. It was at that meeting that Respondent was informed that he could be breaking the law by knowingly hiring undocumented aliens. Respondent claimed that the reference to possible penalties for employing undocumented aliens jogged Nishi's memory of an article he had read a year earlier which stated that persons employing undocumented workers could be fined $500,000.

This asserted business justification for the new policy is belied by a number of factors. Respondent offered no explanation for why he had not adopted a policy against hiring undocumented aliens a year earlier when he apparently first became aware that it might be illegal. It is also unexplained why Respondent adopted the policy of not rehiring workers once they were picked up by the INS, rather than a policy of not hiring undocumented workers in the first place. Moreover, the fact that Respondent did not put the new policy in writing or inform his employees of the change casts substantial doubt on the assertion that the policy existed at all prior to January 1980.

It is also significant that as of August or September 1979, when Respondent allegedly initiated this policy, there were five workers in its employ who had already been picked up at least once by the INS. All five were rehired by Nishi when they returned from Mexico in March 1979 and Respondent did not discharge any of these workers in September 1979, when he claimed to have instituted his new policy.

In light of these contradictions and conflicts in the Respondent's proferred "business justification" for its refusal to rehire the two union activists, the ALO found that on the record in this case it simply was not reasonable to credit this explanation by Respondent. We agree with the ALO's findings in this regard and find that Respondent had no policy against rehiring of undocumented workers who had been picked up by INS until January 1980. We further find that Respondent created such a policy, not for the purpose of complying with the law prohibiting the hire of undocumented workers, but for the purpose of discouraging agricultural employees from engaging in protected union activity.

Applying the Wright Line test to this case, we conclude that Respondent has failed to present any legitimate business justification and therefore failed to show that the discriminatees would have been refused rehire even absent any union activity. We therefore conclude that Respondent, in refusing to rehire Bernal and Batres, violated Labor Code section 1153(c) and (a).

* * *

Question 6. What was Nishi's defense to the allegation that he discriminatively refused to rehire two workers? What was the crucial element of proof that belied his defense?

F. BAD FAITH BARGAINING

ALRA §1153(e) makes it an unfair labor practice for an employer to fail to bargain in good-faith with its employee's certified bargaining representative, and ALRA §1154(c) requires a certified union to bargain in good-faith with the employer. ALRA§1155.2(a) defines good-faith bargaining:

> . . . to bargain collectively in good faith is the performance of the mutual obligation of the agricultural employer and the representative of the agricultural employees to meet at reasonable times and confer in good faith with respect to wages, hours, and other terms and conditions of employment, or the negotiation of an agreement, or any questions arising thereunder, and the execution of a written contract incorporating any agreement reached if requested by either party, but such obligation does not compel either party to agree to a proposal or require the making of a concession.

"Surfacing bargaining" is one form of bad-faith bargaining. Unlike a technical refusal to bargain or a unilateral change in wages or working conditions during bargaining, surface bargaining cannot be determined objectively from the facts. Instead, the Board asks if the accused party was just going through the motions of bargaining while attempting to prevent agreement. The ALRB cannot require either party to make concessions or to agree to specific proposals, but it does consider the bargaining posture and bargaining tactics as a whole to determine if the party has performed its duty to bargain in good-faith. A good discussion of surface bargaining follows in <u>Montebello Rose</u>.

Case 7. Surface Bargaining and Impasse

MONTEBELLO ROSE/MOUNT ARBOR NURSERIES
(1979) 5 ALRB 64

. . .

. . . We are called upon to determine whether Montebello and Mount Arbor have fulfilled their obligation to bargain in good faith with the UFW.

Montebello and Mount Arbor met several times with the UFW, exchanged and discussed proposals, and reached agreement in some areas. Our inquiry does not end there, however, for an employer cannot fulfill its obligation to bargain in good faith merely by meeting with the certified representative of its employees. "Collective bargaining . . . is not simply an occasion for purely formal meetings between management and labor while each maintains an attitude of take it or leave it; it presupposes a desire to reach ultimate agreement, to enter into a collective bargaining agreement" and a "serious intent to adjust differences and to reach an acceptable common ground." (Citation omitted.) We must decide whether Montebello and Mount Arbor negotiated in a good-faith attempt to reach a collective bargaining agreement with the UFW or, instead, conducted negotiations "as a kind of charade or sham," for such a "[s]ophisticated pretense in the form of apparent bargaining . . . will not satisfy a party's duty under the Act." (Citation omitted.)

Our task is a difficult one. We must judge whether Respondents bargained in good faith by examining the totality of the circumstances including the parties' conduct and statements at and away from the bargaining table. In so doing, we must treat the facts as an interrelated whole, for while some conduct standing alone may constitute a per se violation of the Act, other conduct, innocuous in and of itself, may support an inference of bad faith when examined in light of all the evidence. (Citation omitted).

. . .

Period 1: December 3, 1975 - May 12, 1976

On December 3, 1975, the Board issued two certifications whereby it certified the UFW as the collective bargaining representative of the agricultural employees of both Mount Arbor and Montebello. On December 6, Cesar Chavez, the president of the UFW, wrote to each company requesting the commencement of negotiations and certain information relevant to collective bargaining.

Jay Jory, Mount Arbor's negotiator, and Sylvan Schnaittacher, the UFW's negotiator, began to meet in mid-December. At the first meeting, Jory and Schnaittacher discussed the UFW's initial proposal and continued to do so at meetings on December 30, 1975, and January 15, 1976. By the January 15 meeting, Schnaittacher was emphasizing the UFW's desire for a counterproposal from Mount Arbor. On January 23, 1976, Mount Arbor presented counterproposals on five of the 41 articles contained in the UFW's first proposal. Schnaittacher protested Jory's failure to provide a complete counterproposal. Despite repeated requests, Mount Arbor did not present its first complete counterproposal until March 19, 1976. We agree with the ALO that Mount Arbor's failure to present a complete counterproposal until March 19 and its submission, on that date, of a counterproposal calculated to disrupt the bargaining process evidenced bad-faith bargaining. . . .

We also conclude that, in April 1976, Mount Arbor further evidenced its surface bargaining by changing the wages of its budders without prior notice to or bargaining with the UFW about the change. Prior to the 1976 budding season, budders received a bonus if they experienced a 90 percent success rate in their work. In April 1976, Mount Arbor added the bonus to the budders' regular wages, thereby eliminating the 90 percent success rate requirement. This conduct violated §1153(e) and (a) because it constituted a unilateral change; an employer may not unilaterally alter the wages or working conditions of its employees but must, instead, notify and bargain with the certified collective bargaining representative prior to instituting the change. (Citation omitted.)

Initially, Montebello failed to answer the UFW's December 6 letter requesting relevant information and the commencement of bargaining. It was not until the UFW threatened legal action on December 30 that Montebello's representative, William Callan, met with Schnaittacher. Montebello and the UFW held negotiating sessions and discussed the union's initial proposal on January 21 and February 4, 1976. At the February 4 meeting, Callan agreed to submit a counter-proposal. However, Callan never again contacted the UFW and the UFW failed in its repeated attempts to reach Callan. Finally, on April 21, 1976, Jory wrote to the UFW stating that Montebello had retained him to handle its labor negotiations. He stated that Montebello desired to negotiate jointly with Mount Arbor and that he expected Montebello to adopt all of Mount Arbor's bargaining positions.

Thus, at least between February and April 1976, Montebello failed to discharge its duty to provide a representative who was available to meet with the UFW at reasonable times and with reasonable regularity (citations omitted), and to provide the UFW with a counterproposal in a reasonably diligent manner. (Citation omitted.) This conduct is evidence of Montebello's bad-faith bargaining.

...

Period 2: May 13, 1976 - February 2, 1977

On May 13, 1976, the parties conducted their first joint negotiation session. Jory represented both Respondents and Dolores Huerta assumed the collective bargaining responsibilities for the UFW. Between May 13 and June 10, the parties met several times, exchanged proposals and reached agreement on a variety of items. They made substantial progress toward resolving their differences in key areas such as hiring and seniority. On other issues, notably those contained in the economic package, little discussion took place although the parties did exchange proposals. The parties discussed at length the issue of union security but were in substantial disagreement over dues checkoff and discharge for loss of good standing in the Union.

At the May 26 meeting, the parties targeted May 24 [sic] as the date by which they expected to reach agreement on all outstanding issues. Meetings were scheduled throughout the month of June including June 16, 17, 22 and 23. These last meetings, however, were not held. At the June 10 meeting, Respondents abruptly announced that the parties were at impasse. They presented their "final offer" and urged Huerta to put it before the membership for a ratification vote. Huerta rejected the proposal. Jory stated that future negotiations would be pointless and Respondents' representatives left the meeting.

During the ensuing months, the parties exchanged several letters but did not resume substantive negotiations. While Jory repeatedly requested a "constructive response" to Respondents' final offer, the UFW stated that it would not accept the take-it-or-leave-it proposal. On January 29, 1977, however, the UFW wrote to Jory and unconditionally requested the resumption of negotiations.

In the fall of 1976, Mount Arbor raised its employees' wages to the levels it had previously offered to the UFW during negotiations. Montebello raised its employees' wages in the fall of 1976 to a level above what it had previously offered to the UFW. The UFW was not notified or consulted in advance of either of these wage increases. Also, following the Board's denial of the UFW's request for an extension of certification, Jory wrote to both Mount Arbor and Montebello suggesting that they consider "preparing communications to the employees which, while making no promises, urge them now to give the Company a chance in view of the fact that the Union fulfilled few, if any, of its promises in a year's time." Although Mount Arbor did not follow this advice, Richard Barwick, General Manager of Montebello, informed his employees that the UFW had failed to obtain an extension of certification and urged them to give the company a chance to meet the employees' desires without the participation of the Union. This action was taken prior to the expiration of the certification year. . . .

...

Impasse

We turn now to the issue of whether the parties were at impasse on June 10, 1976, when Respondents presented their final offer and declared that further negotiations would be pointless. A bona fide impasse is reached when the parties to collective bargaining negotiations are unable to reach agreement despite their best, good faith efforts to do so.

> Whether a bargaining impasse exists is a matter of judgment. The bargaining history, the good faith of the parties in negotiations, the length of the negotiations, the importance of the issue or issues as to which there is disagreement, the contemporaneous understanding of the parties as to the state of negotiations are all relevant factors to be considered in deciding whether an impasse in bargaining exists. (Citation omitted.)

Our examination of the record, in light of the above-listed factors, convinces us that the ALO properly found that the parties were not at impasse and that Respondents' premature declaration of impasse was indicative of its intention to frustrate negotiations and avoid signing a contract with the UFW.

As a general rule, contract negotiations are not at impasse if the parties still have room for movement on major contract items, even if the parties are deadlocked in some areas. (Citations omitted.) Continued negotiations in areas of concern where there is still room for movement may serve to loosen the deadlock in other areas.

We find that, as of June 10, 1976, the parties had room for movement on several important issues such as hiring, seniority, wages and pensions, and that impasse had therefore not been reached. On several of these issues, the parties had been able to narrow their differences in previous meetings. Had bargaining continued, further movement and ultimate agreement on many points could very likely have been accomplished.

For example, between May 13 and June 10, 1976, the parties made substantial progress toward agreement on contract language covering hiring and seniority despite their widely diverging views on the topic. Each party compromised on several points and, on June 9, the UFW presented new proposed language which incorporated much of the discussion of the previous weeks. Although the parties were still making progress in negotiations, Respondents precluded resolution of their remaining differences by breaking off negotiations two weeks before the parties' own target date for reaching a complete agreement.

On other issues, particularly economic items, Respondents first declared an impasse and then presented their "final offer" before the parties fully discussed their differences or explored possible areas of compromise. For example, Respondents presented their final wage proposal prior to any in-depth discussion on wages. On March 19, 1976, Respondents proposed keeping wages at their then-current levels. On May 13, 1976, Respondents offered a modest wage increase and the UFW responded with its own wage proposal the following day. Between that time and Respondents' action on June, little discussion of wages took place. The Respondents' premature declaration of impasse aborted the negotiating process long before the possibilities for movement and agreement on economic issues were adequately explored.

Respondents, citing <u>Television & Radio Artists v. NLRB</u> (citation omitted), assert, however, that the deadlock on issues such as union security rendered continued negotiations pointless notwithstanding any movement which might have been available in other areas. As of May 13, 1976, the parties were unable to agree on two key aspects of the proposed union security clause: dues checkoff and discharge for failure to maintain good standing. As to dues checkoff, the UFW desired a contract provision whereby the Employer would automatically deduct employees' union dues from their paychecks. The UFW explained that it had used other methods of dues collection in the past and found them to be inadequate. Respondents opposed the inclusion of a dues check-off provision in the contract ostensibly because of the clerical costs. Jory asserted that the extra work would require Respondents to hire additional clerical personnel and that they were unwilling to incur that expense. On good standing, the UFW proposed a clause which would require employees, as a condition of continued employment, to maintain membership in good standing in the UFW. Respondents opposed such a provision to the extent it allowed the UFW to require Respondents to discharge an employee for conduct other than the failure to pay dues and initiation fees. Respondents stated that they maintained this position because the proposed clause would be illegal under the National Labor Relations Act... and because they desired to protect their employees from any arbitrary action by the UFW.

. . .

... [W]e agree with the ALO that the deadlock on the issue of union security resulted from Respondents' bad-faith bargaining posture. Therefore, such a deadlock may not be considered the basis of a legally cognizable impasse. (Citation omitted.) It is a basic principle of both the Agricultural Labor Relations Act and the National Labor Relations Act that:

> The obligation of the employer to bargain in good faith does not require the yielding of positions fairly maintained. It does not permit the Board, under the guise of finding of bad faith, to require the employer to contract in a way the Board might deem proper. (Citation omitted.)

It is equally true, however, that:

> ... if the Board is not to be blinded by empty talk and by the mere surface motions of collective bargaining, it must take some cognizance of the reasonableness of the positions taken by an employer in the course of bargaining negotiations. (Citation omitted.)

Our examination of the positions taken by Respondents on the issue of union security convinces us that Respondents' conduct was not consistent with good-faith bargaining.

We find that Respondents displayed an unwillingness to bargain in good faith about a dues check-off provision because their professed reason for opposing and refusing to compromise on the provision, i.e., the cost of such a system, was pretextual. The UFW was thereby prevented from attempting to address Respondents' true concerns. "Good-faith bargaining necessarily requires that claims made by either bargainer should be honest claims." (Citation omitted.) We believe that the cost factor was not the true reason for Respondents' unwavering opposition to the provision because neither Respondents made any serious effort to estimate the amount of additional work required by a dues check-off system despite the UFW's arguments that the additional burden was not great. Richard Barwick and John Camp, the General Managers of Montebello and Mount Arbor, respectively, and the company representatives responsible for the negotiations, did not substantively discuss the issue with their office staffs, despite their lack of experience in the area. Respondents never explored possible compromises with the UFW to cut down the total amount of paperwork, e.g., by eliminating other paperwork requirements of the contract proposal, or by introducing other cost-cutting measures which could have made the acceptance of a dues check-off provision more attractive. Thus, Respondents insisted that they would not accept a dues check-off provision because of the added clerical burden without having made any effort to determine what the burden would be. Respondents' arbitrary and unyielding rejection of the UFW's dues check-off proposal is thus revealed not as an honestly-held concern but as a method by which to frustrate negotiations and avoid signing a contract.

Respondents' unwavering opposition to the proposed good standing provision is also inconsistent with the duty to bargain in good faith. While an employer may certainly maintain its bargaining positions to the point of impasse, it is an indication of bad faith for an employer to advance "patently improbable" justifications for its stance. Such conduct prevents the other party from seeking possible areas of compromise. (Citations omitted.) Respondents' concern that the proposed good standing provision would not be lawful under the National Labor Relations Act is patently improbable because it has little if any relevance to the negotiations between

Respondents and the UFW; those negotiations are not controlled by the federal labor law. The lack of any logical relationship between the stated concern and the negotiations leads us to conclude that Respondents' justification was pretextual, i.e., a ploy to frustrate negotiations rather than an honestly-held concern. Respondents' second justification for their position, their desire to protect their employees from arbitrary action on the part of the UFW, is equally infirm. It demonstrates a failure to accept a basic principle of the Agricultural Labor Relations Act: the certified collective bargaining representative is the <u>exclusive</u> representative of the employees, and the employer may not assume that role. Respondents' position, viewed in conjunction with their overall conduct, demonstrates a rejection of the Union's role in collective bargaining and, therefore, a rejection of the principle of collective bargaining itself. "Conduct reflecting a rejection of the principle of collective bargaining or an underlying purpose to bypass or undermine the union, in the Board's view, manifests the absence of a genuine desire to compromise differences and to reach agreement tin the manner the Act commands." (Citation omitted.).

Wage Increases

We turn next to the issue of Respondents' wage increases in the fall of 1976. As stated above, Mount Arbor raised its employees' wages to the level it previously offered to the UFW; Montebello raised the wages of its employees to a level above that previously offered to the UFW. Neither Respondent notified or consulted with the UFW prior to instituting these changes. We conclude that this conduct constituted a per se violation of Section 1153(e) and (a). An employer may not by-pass the certified collective bargaining representative of its agricultural employees by unilaterally instituting changes in wages or other working conditions. (Citation omitted.)

Respondents argue that the wage increases were lawful because they were instituted following the occurrence of an impasse in negotiations. While an employer "acquires a limited right to fix [wages and working conditions] unilaterally" after bargaining to a bona fide impasse, (citation omitted), Respondents may not justify their wage increases on that basis in this case. As we previously found, the parties did not bargain to a bona fide impasse and, therefore, Respondents were not entitled to unilaterally change their employees' wages. (Citation omitted.) Furthermore, even after impasse, an employer may change wages and working conditions only within the confines of its prior offers to the union: (citation omitted). Thus, even if a bona fide impasse had occurred, Montebello's wage increase would be unlawful as Montebello raised wages to a level above that which it previously offered to the UFW.

[The Board then discussed the effect of the employers' communications with the employees urging them to "give the company a chance. . . ." The Board concluded that this conduct also violated the duty to bargain in good faith.]

* * *

Question 7a. Did Montebello and Mount Arbor commit ULPs by unilaterally changing wages? Why or why not?

Question 7b. Assume that the Board found that the employers gave pre-textual reasons for opposing the dues check-off provision, and that this was bad-faith bargaining. What if the employers had given no reason to oppose dues check-off?

Case 8. Negligence Versus Bad-Faith Bargaining

SAM ANDREWS' SONS, INC.
(1982) 8 ALRB No. 64

. . .

Facts

Negotiations began in January 1979 between the United Farm Workers of America, AFL-CIO (UFW) and Respondent (Sam Andrews), with Paul Chavez as the UFW negotiator and Tom Nassif as negotiator for Respondent. A number of language proposals had been exchanged by the time that Ann Smith took over negotiating for the UFW in September 1979. However, little progress had been made.

On November 5, 1979, the UFW submitted its first complete proposal to Respondent. The Company responded with a complete counter-proposal on November 7, 1979. Both proposals drew heavily from the Sun Harvest contract negotiated earlier that year between the UFW and Sun Harvest.

. . .

On February 12, 1980, Don Andrews, the company principal responsible for labor relations, dictated a cassette tape, which was transcribed on February 13 by Nassif. Noting gaps or long pauses on the tape itself, Nassif contacted Don Andrews and informed him of the "gaps." Nassif then gave the transcription to Don Andrews to check and make necessary corrections. Andrews reviewed and corrected his copy of the transcript, which had indicated acceptance, rather than rejection of Sun Harvest Articles 44-47. Andrews also indicated that there were no gaps in the tape. Andrews sent Nassif a photocopy of the corrected transcription; however, Nassif did not check the returned transcript for corrections.

Nassif testified that he used the original uncorrected transcription, along with Andrews' comments and previous proposals and agreements, to compose a March 21, 1980, letter to the UFW. This letter constituted Respondents' counterproposal to the UFW's January 24, 1980 proposal regarding the Sun Harvest contract. This counterproposal was mailed to the UFW on March 21, 1980, with a copy to Andrews, incorporating approval of Articles 44-47 of the Sun Harvest contract.

The proposal was received by the Union on March 24, 1980, and answered on March 25, 1980, by Ann Smith who expressed encouragement at the movement in the Company's bargaining position and specifically mentioned the movement on Sun Harvest Articles 45 and 46. Ann Smith testified that she believed the March 21 letter to be the latest Company proposal because it listed the Company's position on various contract terms, including changes in the Company's position. She testified that the union believed the Company was changing positions on Articles 44-47 with the intention of moving the bargaining forward. Tom Nassif testified that he too saw it as a major move by the Company, that it was a good sign in that the parties had been stalled for some time.

Don Andrews received all this correspondence, but after reading the first paragraph of the March 21 letter, put it in a file and did not make any detailed analysis of it. . . .

At the next negotiating session on April 15, 1980, Smith went through the Company's March 21, 1980 proposal, item by item, including Articles 44, 45, and 46. The Union, relying on the Company movement in the March 21 proposal, changed its position by offering a 10-cent cotton differential, a significant concession since the Union had, to this point, insisted on an equal wage scale in all job operations in all crops. Although both Nassif and Andrews were present, neither made any comment at the time regarding any error. Andrews took extensive notes during this session and placed question marks next to Articles 45-47 of the Sun Harvest contract.

. . .

Nassif testified that when he and Andrews went back to Nassif's office after the negotiating session, Andrews told him that Andrews could not understand why Smith believed they had reached agreement on those articles. Nassif told Andrews that there was agreement on the basis of the tape and transcription and that Nassif had made the March 21 proposal, accepting those proposals. They then went and listened to the tape in Nassif's office and Nassif then realized that there was a mistake.

Nassif called Smith that afternoon, apologizing and informing her of the error and how it had been made. Smith could not believe that such a mistake was possible, given all the correspondence and discussion of these issues. Smith further stated that this mistake was unacceptable to the Union and would result in an unfair labor practice. Nassif immediately offered to allow the union to withdraw their concession on the cotton differential that was made in reliance upon the mistake. Smith reiterated that the Union would hold the Company to their position, pointing out that Andrews had been sitting there at the meeting as she specifically listed the articles without batting an eye or indicating any problem, and it was just too hard to believe. Nassif talked with Smith a second time that day on the phone and told her that a mistake this obvious could have only happened by inadvertence and was not intentional. Smith then met with members of the employee negotiating committee, and committee members were angry and disappointed with the "illusion" of progress.

. . .

Discussion

The UFW argues in its exceptions that there was no mistake, and further, that the retraction was made to prevent agreement, not to correct an error. Respondent argues that the retraction was based on an honest mistake, and that, since there was good cause for retraction, there is no evidence of bad faith.

It is well established that withdrawal of tentative agreement on bargaining proposals, without good cause, is evidence of bad-faith bargaining. (Citations omitted.) However, the record in this case establishes that the miscommunication between Respondent and its negotiator caused a genuine mistake, and was not as the UFW contends, an effort to intentionally mislead the Union. We therefore find, in the isolated context of Respondent's March 21 proposal, that Respondent had good cause for withdrawing its proposal and did not renege on a tentative agreement. We are not persuaded, however, that Respondent's mistaken proposal of March 21, 1980, was consistent with the duty to bargain in good faith.

A company's good faith may be tested by considering whether it would have acted in a similar manner in the usual conduct of its business negotiations. (Citation omitted.) That is, a company must treat the bargaining obligation as seriously as it would any other business transaction. The failure to devote sufficient time and attention to the bargaining obligation has been found to be disruptive and in derogation of the collective bargaining process. (Citations omitted.)

In the instant case, it is evident that there has been negligence on the part of both principal Don Andrews and negotiator Tom Nassif. Respondent demonstrated a lack of seriousness toward the negotiations by failing to check its own proposals for accuracy and by paying no attention to communications between its negotiator and the UFW. We find that Respondent's inattention to its own communications with the Union evidences a lack of good faith and sheds doubt on the seriousness of Respondent's desire to reach agreement. . . . We do not find, however, that this incident, which is only a part of the long and complex bargaining history between the parties, is sufficient in itself to prove that Respondent was bargaining in bad faith overall. We have stated that a finding regarding bad faith bargaining must be based on the totality of the circumstances of the negotiations. (Citation omitted.) This case presents too few circumstances to support such a finding.

Order

By authority of Labor Code section 1160.3 the Agricultural Labor Relations Board hereby orders that the complaint herein be, and it hereby is, dismissed in its entirety.

* * *

Question 8a. Was Andrews' behavior negligence in bargaining or bad-faith bargaining? Why?

Question 8b. It is sometimes said that the Board gets the parties into the negotiating room and then closes the door. Since the Board does not participate in bargaining, why should the parties' negotiating proposals be admitted as evidence in a Board hearing? Do you think the terms of each party's bargaining proposals should be admitted as evidence in a hearing to decide a bad-faith bargaining charge? Why?

Case 9: Lockout

A lockout occurs when an employer reduces work or lays off employees in anticipation of a labor dispute. The lockout can be a disruptive economic weapon, and its use is limited by the ALRA. Determining what actions constitute a lockout can be difficult in agricultural labor relations because of the nature of agriculture.

WEST FOODS
(1985) 11 ALRB No. 17

. . .

From July 1981[4] through the conclusion of the hearing in this case (April 1982), West Foods and the UFW fruitlessly negotiated towards a collective bargaining agreement for Respondent's [Employer's] Ventura operations. During this period, both parties employed their economic weapons: West Foods began an allegedly unlawful phasedown of its operations on July 15, 1981, and the UFW called a strike on November 19. West Foods is also alleged to have engaged in a variety of acts constituting bad faith bargaining, a number of which the ALJ found. Respondent excepted to each of these conclusions. We first turn to consideration of the legality of Respondent's phasedown of its operations which was found by the ALJ to constitute an unlawful lockout.

Lockout

Respondent grows and packs mushrooms at its Ventura plant on a year-round basis. The mushrooms are grown in cycles so that at any given time some mushrooms are ready for harvest while others are at various earlier stages of growth.

The UFW was certified as the collective bargaining representative of Respondent's agricultural employees on December 4, 1975. Since then, the parties have negotiated two contracts covering the Ventura unit, one from September 6, 1976 to September 6, 1978 and the other from September 6, 1978 to September 6, 1981. During the same period, Respondent also had a contractual relationship with the UFW at another unit in Soquel, California, the history of which also figures in the matter before us. The negotiations which resulted in the 1976-1978 Ventura contract were highlighted by a strike which, although it resulted in a contract, entailed considerable economic loss to Respondent. In the following year's (1977) Soquel negotiations a contract was achieved without economic action.

With its 1976 experience clearly in mind, Respondent decided to approach the Ventura negotiations in 1978 with an eye to avoiding a strike. Company officials approached Cesar Chavez, president of the UFW, to inform him that the Company's fragile financial condition made it necessary to avoid a strike. They told Chavez that if negotiations did not conclude early, the Company would have to consider putting into effect "a crop protection program" under which operations would be phased down so that the Company would be completely shut down by the expiration date of the contract. When, in response, the Union suggested extending the contract with its "no-strike, no lockout"[5] provision, Respondent replied there was ample time to reach an agreement before the expiration date. Negotiations commenced in mid-July and were in mid-stream when, on August 18, 1978, Respondent began its phasedown by ceasing to prepare compost, which is the first step in its production cycle. A new contract was reached on August 24, 1978.

. . .

The Ventura negotiations which are the subject of this case took place not only against this backdrop, but also against a background of hostility which was apparently peculiar to the Ventura unit itself. On the one hand, the Union perceived Respondent as having undermined the contract, and, on the other, Respondent perceived the Union as willing to attack it economically. As the ALJ has detailed, each side received, relied upon, and, in the case of [Employer], solicited reports of the other side's hostility and willingness to resort to their respective weapons and counter-measures. As was true of the approach to the 1978 Ventura and 1979 Soquel negotiations, Respondent again determined to press for an early settlement and to make the Union aware that, should the parties fail to achieve one, Respondent would resort to a phasedown. Unlike the earlier negotiations, however, when Respondent threatened resort to crop protection only a few weeks or a few days before the contract was due to expire, Respondent decided to implement the crop protection program much earlier than in any of the previous negotiations. According to Respondent's witnesses, the added lead time for a phasedown in 1981 was made necessary because Respondent's high production levels made it all the more critical to avoid a strike. . . .

In May 1981, Respondent's negotiators met with Chavez at La Paz, California to inform him of the plan to phasedown operations if an early settlement was not reached. Chavez told Company representatives that the Union had not yet appointed its negotiator for the Ventura region and that he could not discuss the matter. On June 17, 1981, UFW negotiator Roberto de la Cruz and the Ranch Committee met with George Horne, the Company's negotiator, at Respondent's offices. Horne told the Union that the Company wanted a contract before July 15, 1981 or a 30-day extension, or it would shut down after that date. As the Company had refused to consider an extension during the 1978 Ventura negotiations, de la Cruz now refused to consider one on the grounds that there still was plenty of time to negotiate. Two days later, the Union submitted a request for information and scheduled a negotiation session for July 6, 1981, a little more than a week before the phasedown was scheduled to commence. The Company responded on June 25, 1981 by providing some information, by offering to make other information available and, finally, by telling the Union it already possessed still other information.

[Between July 6 and July 16 the parties exchanged proposals. The company requested an extension three times, and the union declined three times. On July 16 the company implemented the "phasedown."] Implementation of the program resulted in the demotions, reassignment of work and intermittent layoff of employees which are described in the accompanying ALJ Decision.

Treating the phasedown as a "lockout," the ALJ found that it was violative of Labor Code section 1153(e) because it was an integral part of Respondent's bargaining strategy, in bad faith and inherently prejudicial to employee interests. Respondent vigorously objects to this conclusion, arguing that the phasedown was not a "lockout" but a lawful "defensive" measure undertaken to protect its business. We affirm the conclusion of the ALJ that Respondent unlawfully "locked out" its employees in violation of Labor Code section 1153(e) prior to the expiration date of the contract; while we adopt the ALJ's analysis as an additional basis for finding a violation of section 1153(e), we rely upon our own analysis of the intent of the Legislature in enacting section 1155.3(e). Since we affirm his finding that Respondent was guilty of overall bad faith bargaining, we also affirm the ALJ's conclusion that so far as the lockout continued past the expiration date of the contract, it was also violative of the Act.

Because the shutdown in this case took place within the 60-day period preceding expiration of a collective bargaining agreement, the starting point for analysis must be Labor Code section 1155.3(a) which states:

1155.3(a) Where there is in effect a collective-bargaining contract covering agricultural employees, the duty to bargain collectively shall also mean that no party to such contract shall terminate or modify such contract, unless the party desiring such termination or modification does all of the following:

(1) Serves a written notice upon the other party to the contract of the proposed termination or modification not less than 60 days prior to the expiration date thereof, or, in the event such contract contains no expiration date, 60 days prior to the time it is proposed to make such termination or modification.

* * *

(4) Continues in full force and effect, without resorting to strike or lockout, all the terms and conditions of the existing contract, for a period of 60 days

after such notice is given, or until the expiration date of such contract, whichever occurs later.

On its face, then, section 1155.3(a) clearly proscribes "lockouts" within the very period in which Respondent implemented its phasedown. Accordingly, it follows that if the "phasedown" was a "lockout" within the meaning of section 1155.3(a), Respondent's action was, by definition, a refusal to bargain. However, (citations omitted) Respondent argues that the phasedown of its operations from July 15, 1981 forward was not a "lockout," but was, instead, a lawful measure to protect itself from the hardship of a strike. . . .

Fewer concepts in labor law have created such definitional problems as that of the lockout. . . . Although the term is utilized in the NLRA, it is nowhere defined by that Act . . . nor has it been consistently used by the National Labor Relations Board. (Citation omitted.) Nevertheless, because it is clear that under the circumstances of this case, a "lockout" will contravene section 1155.3(a)(4), determination of what that section aims to prevent will outline the area of our inquiry into the lawfulness of Respondent's actions.

Section 1155.3(a)(4) is the analog to section 8(d)(4) of the Taft-Hartley Act. (Citations omitted.) This section was added by Congress to the NLRA:

> . . . to assure that, once parties have stabilized their bargaining relationship by entering into a contract, the stability achieved will not be placed in jeopardy by strikes or lockouts. It is for this reason that the section provides for a waiting period before strike or lockout action by the parties. Clearly, Congress was interested in establishing an orderly procedure for contract negotiations and in preventing the industrial unrest that is the natural consequence of the failure of the parties to abide by their collective bargaining agreement. (Citation omitted.)

Accordingly, section 8(d) "seeks, during this natural renegotiation period, to relieve the parties from the economic pressure of a strike or lockout in relation to the subject of negotiation. (Citation omitted.) (Emphasis added.). . . And if it is true that a lockout in aid of an employer's bargaining strategy during the "cooling off" period violates 1153(e), it must be all the more true that a lockout which is the centerpiece of an overall strategy of bad faith bargaining by an employer, as the ALJ found Respondent's bargaining to be, will also violate 1153(e).

. . .

As was true in previous contract negotiations between Respondent and the Union, the phasedown and the July 15, 1981 date for its implementation was formulated well in advance of negotiations. While one of the stated purposes of the crop protection plan was to avoid the effects of a potential strike occurring after the expiration of the contract, other stated purposes were to force the UFW, prior to the expiration of the contract, to agree to a contract, to extend the current contract beyond its expiration date, or to give a 60-day notice of any future strike. As the crop protection program constituted action designed to put economic pressure on the Union to agree to an early contract or to make concessions regarding the Company's proposals, it falls within the meaning of the term "lockout." Inasmuch as such economic pressure was applied on the Union during negotiations ... after notice to terminate or modify the contract was given, but before expiration of the contract, the implementation of the crop protection program was at odds with the purpose of section 1155.3(a)(4). ...

... Accordingly, to the extent that [Link-Belt (citation omitted] is at all instructive for present purposes, it makes the inquiry into Respondent's motive, which we undertake in this case, critical. Similarly, although Betts Cadillac-Olds, Inc., (citation omitted), did not involve accommodating the tensions between 8(d)(4) and the right of an employer to defend itself, the Trial Examiner in that case also specifically focused his inquiry on whether the motive of the Respondents was "defensive" or "offensive." (Citation omitted.) In American Brake Shoe v. NLRB, (citation omitted), a case which did involve "defensive" action during the "cooling off" period, the Board conceded that the company "was motivated solely by foreseeable operative and economic difficulties as a result of its apprehension of a possible strike." ...

We recognize that our inquiry into the interplay between the proscriptions of 1155.3(a)(4) and an employer's economic defense revitalizes the distinction between offensive and defensive lockouts which, in the wake of American Shipbuilding v. NLRB (citation omitted) and NLRB v. Brown (citation omitted), no longer applies outside the waiting period. However, in view of the tension between the clear purpose of the "cooling off" period and the continued vitality of the economic defense, there is no way to avoid such a result. Our task, then, is to accommodate the tension between these two competing interests. In seeking to strike an appropriate balance, we are aware that:

[a] loose application of the economic defense to lockouts within the cooling-off period would involve the risk of eroding the statutory moratorium. On the other hand, an absolute proscription of employer-initiated shutdowns during the moratorium period would involve the risk of inflicting extraordinary losses on employers including losses from physical damage to plant and raw materials, losses, which [in the past] have produced the best case for economically privileged lockouts. (Citations omitted.)

Our task is all the more delicate because, unlike the NLRB which regulates a broad range of industries not all of which could plausibly claim the need to protect a perishable product, almost all of the industry we regulate produces quickly perishable commodities. We do not believe the legislature, in incorporating the "cooling off" period into our statute, intended to permit it to be easily ignored by the merest claim of economic necessity. Since crops cannot be shut down on a single day, a too facile application of the economic defense would permit a phasedown of operations by an employer at the very outset of the growing cycle, regardless of when in the production cycle the contract expires. This result would render section 1155.3(a) meaningless since in almost all instances, an employer could shut down well in advance of the expiration of its contract if that expiration date encompassed any point in the production cycle. Instead, the intent of the Legislature is clear: parties should have the statutorily prescribed period of time in which to use good faith efforts to negotiate a new contract without resort to economic weapons to force concessions from the other party. . . .

Our conclusion that Respondent's motives were not "defensive" is reinforced by our finding that Respondent's fear of an imminent strike was not reasonable.
. . .

In this case, the only strike Respondent had experienced was in 1976. Since then Respondent had negotiated other contracts with the Union without any threat of a strike: the 1977 Soquel agreement and the 1978 Ventura agreement. The ALJ concluded that there was no record evidence of work stoppages during the term of the 1978-1981 Ventura contract which would support Respondent's fear that a strike was imminent in 1981. The ALJ also rejected Respondent's contention that UFW members, agents and representatives threatened a strike. Indeed, to the extent either party threatened economic action, it was Respondent which had incorporated not only the threat, but also the use of economic action into its negotiating position. Although the Union obviously resisted giving Respondent any assurance that it would seek to achieve an early contract, for its part Respondent showed no alacrity in providing the information the Union requested to formulate its proposals. We find it odd for Respondent to make the Union's lack of diligence a fault where it plainly showed itself to be less than diligent.

Respondent's contention that the UFW refused to give any assurances that it would not strike is false. The record is replete with testimony from UFW negotiators, as well as Company negotiators Jim Kahl and George Horne, that the Union repeatedly told the Company at the negotiation sessions in July and August that it would not strike.

Having found that Respondent did not have a reasonable fear that a strike was imminent, and that the crop protection program was in fact economic action designed to apply pressure for contractual concessions upon the Union during the time period specified in section 1155.3(a), we conclude that the crop protection program was a lockout prohibited by section 1155.3(a) and hence unlawful. . . .

. . .

* * *

<u>Question 9</u>. What could the company have done to make the phasedown legal?

G. INTERNAL UNION AFFAIRS

ALRA §1154(a)(1) states that "it shall be an unfair labor practice for a labor organization or its agents to restrain or coerce . . . agricultural employees in the exercise of the rights guaranteed in §1152." In UFW (Marcel Jojola, 1980) 6 ALRB No. 58 the ALRB considered the question of whether picketing the homes of strikebreakers was a violation of §1154(a)(1).

Case 10. Residential Picketing

UFW (MARCEL JOJOLA)
(1980) 6 ALRB No. 58

. . .

Facts

This case concerns three separate incidents of residential picketing in March and April 1979, at two homes in Calexico and one in Holtville. The UFW stipulated that the pickets in all three incidents were its agents.

. . .

Residential Picketing

We affirm the ALO's conclusion that the UFW violated §1154(a)(1) by picketing employees' residences in large numbers, shouting obscenities, addressing abusive epithets to the residents and, in the case of the Guerra home, commencing picketing before dawn. Respondent argues that this conduct constituted only peaceful picketing, but we conclude that in the residential settings where it occurred the conduct had a tendency to coerce or restrain agricultural employees in the exercise of protected rights, in violation of Labor Code §1154(a)(1).

The essence of coercion or restraint is that a person is forced, according to the dictates of another and against his or her own judgment and will, to act or to refrain from acting in a certain way. Coercion and restraint attack the autonomy and integrity of the human person. The Act fosters the autonomy and integrity of agricultural employees by protecting their freedom of choice. Section 1152 provides not only that "Employees shall have the right to self-organization, to form, join or assist labor organizations, to bargain collectively through representatives of their own choosing, and to engage in other concerted activities for the purpose of collective bargaining or other mutual aid or protection," but also that employees "shall also have the right to refrain from any or all such activities." . . . This Board cannot condone union conduct violating the freedom of choice that §1152 guarantees.

While picketing is a form of expression and is therefore entitled to protection under the First Amendment to the United States Constitution, both the United States Supreme Court and the California Supreme Court have held that picketing is entitled to less protection than other forms of expression. (Citations omitted.) California's high court has attributed the lower degree of protection for picketing to the "fact that, of itself, picketing (i.e., patrolling a particular locality) has a certain coercive aspect." (Citation omitted.)

The California Supreme Court has recently stated that a union may violate §1154(a)(1) by picketing which obstructs access to a work site to the extent that the picketing "restrains or coerces nonstriking employees in the exercise of their right to refrain from concerted activities guaranteed by §1152." (Citation omitted.) The coercive impact of picketing is likely to be far greater at one's residence than at a work site.

. . .

Our tradition of respect for the domestic sanctuary has created throughout American society an expectation of undisturbed privacy in the home. When the privacy and tranquility of the home are violated by conduct like that which the picketing union agents displayed here, the impact on the resident(s) will inevitably be upsetting and intimidating. In concluding that such conduct is coercive within the meaning of §1154(a)(1), we are recognizing "a connection between preserving the sanctity of the home and protecting the integrity of personality. "S. Hufstedler, The Directions and Misdirections of a Constitutional Right of Privacy (New York 1971), p. 25. We believe that the freedom of choice afforded to employees by §1152 requires that this connection be maintained.

. . .

* * *

Question 10. Would the same picketing as occurred in Jojola be a violation of §1154(a)(1) if it occurred outside a labor camp?

Case 11. Expulsion from the Union

The ALRA grants unions more power to discipline individual members than does the NLRA. However, ALRA §1154(b) states that it is an unfair labor practice for a union:

> To cause or attempt to cause an agricultural employer to discriminate against an employee in violation of subdivision (c) of §1153, or to discriminate against an employee with respect to whom membership in such an organization has been denied or terminated for reasons other than failure to satisfy the membership requirements specified in subdivision (c) of §1153.

The ALRA allows an employer and union to agree that a condition of each worker's employment is membership in good standing in the union if the union is the certified collective bargaining representative of the workers. ALRA §1153(c) provides:

> For the purposes of this chapter membership shall mean the satisfaction of all reasonable terms and conditions uniformly applicable to other members in good standing; provided that such membership shall not be denied or terminated except in compliance with a constitution or bylaws which afford full and fair rights to speech, assembly and equal voting and membership privileges for all members, and which contain adequate procedures to assure due process to members and applicants for membership.

The NLRA differs from the ALRA in that the NLRA allows a union to terminate membership only for failure to pay dues or initiation fees. The ALRA allows termination for a member's failure to satisfy reasonable terms and conditions of union membership.

In UFW (Scarbrough) three workers charged that they were fired in violation of ALRA §1153(c) and §1154(b) for violating one of 33 conditions in the UFW constitution which members must satisfy to remain in good standing.

UFW (SCARBROUGH)
(1982) 8 ALRB No. 103

. . .

In the present case, we are called upon to determine in this matter whether the UFW's constitutional provision requiring all members to honor its strikes and picket lines is a reasonable term or condition of continued membership, and whether the UFW has afforded the Charging Parties due process before terminating their membership in the Union.

Charging Parties were employed as agricultural employees by agricultural employers Mann Packing and Sun Harvest when strikes and picketing began against a number of lettuce growers in January 1979 after strike votes by a majority of the employees at those companies. Although each Charging Party joined the strike initially, each subsequently returned to work while the strike and picketing were still in progress. . . . At the time the Charging Parties returned to work, they were members of the UFW. Also at that time, the UFW Constitution, Article XVIII, Section 4 required that all members honor duly-authorized strikes and refrain from crossing UFW picket lines.

On September 4, 1979, separate collective-bargaining agreements were reached between the UFW and each of the two employers. Each contract contained a union-security clause requiring, as a condition of continued employment, that all employees become and remain members in good standing of the UFW. After the striking employees returned to work, charges were brought by union members against the Charging Parties, alleging that they had violated the union constitution by failing to honor the UFW strikes and picket lines by working for struck employers during the strike. The Charging Parties were thereafter tried by their Ranch Communities, found guilty as charged, and ordered to be expelled from the Union. . . .

. . .

The Reasonableness of the UFW's Membership Requirements

. . .

[The Board first reviewed the legislative history of the ALRA pertaining to §1154(b) and §1153(c).] This history indicates that the Legislature understood and intended that, under §1153(c), agricultural unions would be able to require a good deal more of their members than simply the payment of dues, as under NLRA section 8(a)(3). The determination as to how much more could be required was left to the case-by-case review of this Board. Although the Board's discretion in this area is broad, we are instructed that only union membership requirements which are fair and reasonable may be enforced through expulsion from the union and termination of employment under a union security agreement.

In UFW (Conchola), [1980], 6 ALRB No 16, we held that the union could not require that its members contribute to a fund which was used for contributions to political candidates or for political or ideological purposes that are not germane to collective bargaining. In that case, we distinguished between union expenditures for medical or educational benefits which provide for the general welfare of the membership, and purely political or ideological expenditures which have no direct relationship to the overall purposes that the union could serve through collective bargaining....

In this case, the question is whether the UFW's requirement regarding the honoring of picket lines and strikes is reasonably related to a legitimate union interest that is germane to its role as collective bargaining representative.... We are convinced that the ability to enforce disciplinary rules during a strike is necessary for a union to be able to carry out its duties as the employees' representative....

...

Member McCarthy argues in dissent that since the right of an individual employee to refrain from participating in a strike is protected by §1152 of the Act, this Board may not sanction the punitive loss of employment for exercising that right. Section 1152 reads as follows:

> Employees shall have the right to self-organization, to form, join, or assist labor organizations, to bargain collectively through representatives of their own choosing, and to engage in other concerted activities for the purpose of collective bargaining or other mutual aid or protection, and shall also have the right to refrain from any or all of such activities except to the extent that such right may be affected by an agreement requiring membership in a labor organization as a condition of continued employment as authorized in subdivision (c) of Section 1153. (Emphasis added.)

We are aware that two fundamental purposes of the Act—encouragement of collective activity and protection of individual rights of association—cannot always be neatly reconciled and are, in fact, in direct conflict in this case. However, we believe the underscored language in §1152 and the legislative history of §1153(c) indicate the Legislature's intention that where the union has imposed a membership requirement that is reasonably related to its role in collective bargaining, including its role in exerting economic pressure on the employer, the collective interest of the union and all the striking employee-members supersedes the right of an individual employee to refrain from engaging in union or concerted activity. Moreover, since the Legislature contemplated that termination of employment under a union security agreement could result from loss of good standing, it is not for the Board to say that loss of employment is too harsh a penalty for crossing a picket line during a strike.

Procedural Due Process

Having concluded that the UFW's picket line requirement is reasonable, we must now inquire whether the expulsion procedures of the Union provided adequate due process. As mentioned above, Labor Code §1153(c) states that a labor organization shall not deny or terminate union membership," . . . except in compliance with a [union] constitution or bylaws which . . . contain adequate procedures to assure due process to members and applicants for membership." We interpret that language as requiring unions to provide the basic elements of a fair hearing, i.e., prior notice and an opportunity to prepare and present evidence. This interpretation is supported by both state and federal precedent which states that standards of procedural fastidiousness which apply to criminal or civil litigation in the courts need not be met by internal union procedures. (Citations omitted.) We have reviewed the UFW's constitutional procedures for hearing and resolving charges against union members and find those procedures, on their face, to be adequately fair. . . .

Charging Parties also except to the ALO's failure to find that it was unlawful for the UFW to apply its disciplinary procedures retroactively and his failure to find that . . . the charges against Charging Parties were untimely, based on the UFW's own 60-day limitation. These exceptions are without merit. Charging Parties were voluntary members of the Union at the time they committed the violations and they were therefore subject to its rules at that time. A violation of the UFW constitution was committed by each Charging Party each working day from his return to work until the end of the strike. The internal union charges were clearly filed within 60 days after the last such violation by each Charging Party and were therefore timely.

. . .

The Case of Juan Martinez

[Charging Party Martinez appealed his November 9 expulsion to the National Executive Board (NEB) of the UFW on November 13, which resulted in an automatic stay of the Ranch Community's order of expulsion. On January 2, 1980, Martinez' penalty was reduced by the NEB to a one-year suspension. However, after the UFW gave notice to Mann Packing on January 10 that his membership had been suspended and requested his discharge, Mann Packing discharged him on January 14, 1980.]

We must determine whether the internal union disciplinary proceeding involving Martinez provided the basic elements of a fair hearing required by Labor Code §1153(c). The ALO found that Martinez was denied due process because: (1) the charges against him were vague; (2) one of his trial judges prejudged his guilt; (3) he had no opportunity to cross-examine witnesses against him; and (4) the NEB conducted an insufficient appellate review. As we conclude that Martinez was denied a fair hearing because he had no opportunity to cross-examine witnesses against him, and generally did not understand the trial process, we need not, and do not, reach the other findings of the ALO.

When ranch committee member Rigoberto Perez served Martinez with charges alleging that he had violated the UFW Constitution by crossing the picket line, Perez told Martinez that he had the right to defend himself. Martinez testified that Perez told him he could have a representative at the trial and could present evidence, but Martinez did not recall Perez explaining his right to cross-examine witnesses. Perez explained that the ranch committee was there to serve Martinez and help him in any way it could.

Perez testified that Martinez told him he knew how to defend himself, and Martinez testified that he knew he could present evidence at the trial, but that he had none. However, the transcript of the trial itself, read in light of Martinez' testimony, indicates that Martinez did not have a sufficient understanding of the trial procedure for us to conclude that he was fairly tried.

At the trial, the director of the UFW's field office and a member of the ranch committee read the section of the UFW Constitution covering trial procedures and asked if there were any questions about the procedure. Martinez did not remember hearing the Constitution read, and did not remember being told at the trial that he could present witnesses and ask his accusers questions. Although Perez testified that Martinez asked each witness questions, the trial transcript indicates that Martinez was not asked if he had any questions, and he did not cross-examine any of his accusers.

Martinez testified that he completed only two years of school and could not read or write English or Spanish. While he testified that he knew he could be represented at the trial, his testimony also indicates that he did not know what it meant to be "represented" and did not know the difference between a representative at the trial and a witness on his behalf. At the trial, Martinez asserted his innocence and, although his statements at the trial are somewhat unclear, attempted to assert that he crossed the picket line only once, and he wàs "helping out" rather than working. Martinez told Perez at the trial to go to Growers Exchange and see "if (his) name is there." This suggestion that some work records would support Martinez' version of the events is further supported by Martinez' attempt, after the trial, to give the Union a document from Growers Exchange which indicated that his social security number did not appear in Growers Exchange's payroll records in 1979. Martinez testified that he did not produce this document earlier out of "ignorance." Since Perez told Martinez that the ranch committee was there to help him in any way it could, it is likely that Martinez believed that the ranch committee would check Growers Exchange's payroll records as he requested. Martinez also apparently believed that he could submit a document supporting his defense after the close of the trial.

Given Martinez' lack of understanding of the trial procedure, we cannot find that Martinez could fairly be deprived of his good standing in the Union, and discharged from his job, based on that trial procedure. The ranch committee and ranch community, which had the power to remove Martinez from good standing in the Union, had an obligation to insure that Martinez had a sufficient understanding of his own trial. This obligation it did not meet. Accordingly, we find that the UFW violated section 1154(b) and (a)(1) by terminating Martinez' membership and requesting that Mann Packing Company discharge him, and that Respondent Mann Packing Company violated section 1153(c) and (a) by discharging him.

. . .

* * *

Question 11a. Why did the Board find the UFW requirement that union members must not work at struck farms during the strike to be reasonable?

Question 11b. What if the UFW constitution adopted a rule saying that a UFW member who filed charges against the UFW with the ALRB could be expelled—do you think the ALRB would approve such a rule as reasonable?

Question 11c. Did Martinez get a "fair trial" from the UFW? Why or why not? Did Martinez's employer violate the ALRA for firing him?

H. ANSWERS TO QUESTIONS

Q. 1a. Hansen was the workers' employer and the source of their jobs and wages. Since the workers were economically dependent on Hansen, they may believe that Hansen and not the union determines wages, but the basic concept of collective bargaining is that the employer and union jointly negotiate wages.

Q. 1b. A union may contest a "lost election" in hopes of having the election set aside and holding another election without a 12-month election ban. It is also possible that if the employer's conduct was especially egregious, the employer may be ordered to bargain with the union even though the union "lost" the election. If another union won the election the losing union would want to get the election set aside so the other union would not be certified.

Q. 2. The NLRB decides on a case by case basis whether nonemployee organizers should have access to an employer's property. The ALRB has promulgated regulations allowing limited access in almost all cases. The blanket rule is necessary in agriculture because other channels of communication with farmworkers are generally unavailable. Employees do not arrive or depart on fixed schedules, there are no public areas where employees congregate, employees often have no permanent address or telephones or live in widely spread-out areas, farmworkers speak many different languages and many are illiterate. These factors in combination with seven-day elections mean adequate communication through other channels is ineffective.

Q. 3a. The employer (1) brought the workers together with an attorney to assist in decertification; (2) favored the worker who circulated the petition; and (3) permitted the decertification petition to be circulated at its Christmas party in the presence of supervisors.

Q. 3b. The margin of victory is irrelevant in a case of employer assistance to a decertification campaign. The election is supposed to be an indication of employee choice and the very fact that the employer participated invalidates the election.

Q. 4. Both the Board and the California Supreme Court found that San Clemente was a successor employer to Highland because (1) it produced the same crops; (2) on the same land; (3) in the same manner; (4) the number of workers hired remained approximately the same; and (5) San Clemente acquired Highland's agricultural machinery.

Q. 5. Evidence to support the four elements of proof in this 1153(c) case includes: (1) A discriminatory act: Barba denied his crew the opportunity to purchase their equipment from the company at a discount. (2) Union activities by discriminatees: A majority of the Barba crew attended meetings at the home of a UFW organizer; members of the crew were on a negotiating grievance committee; a majority of the Barba crew wore union buttons and joined in a "union clap" in the field with Barba present; a number attended the pre-election conference; the crew staged a demonstration in support of the UFW; and members of the crew filed an unfair labor practice charge against the company and attended an ALRB hearing. (3) Employer knowledge of the union activities: Barba saw his crew meet with a union organizer in the field and heard them in a "union clap." Baker (the general manager) and Sanchez (the farm labor contractor) saw the pro-UFW demonstration.

Crew members wore union buttons and hats on the job and were seen by management at the pre-election conference. (4) Anti-union motivation or causal connection between the discriminatory act and the union activity: Here the company demonstrated an anti-union stance by opposing the union and increasing wages during the election campaign. Immediately following the campaign it denied the Barba crew the opportunity to purchase equipment. The company also failed to assert any justification which could stand up to scrutiny.

Q. 6. Nishi contended that he discharged Batres and Bernal because he had a policy of not rehiring workers who had been picked up by the INS. His defense was contradicted by the fact that he rehired at least five other workers who had been picked up by the INS after he claimed he had initiated this policy. Therefore, Nishi failed to meet his burden of proving that he would have taken the same action in the absence of the discriminatees' protected union activities.

Q. 7a. Yes. It is a ULP to unilaterally change wages unless there is a bona fide bargaining table impasse. In this case, the ALO concluded that there was no bona fide impasse on June 10, 1976, so the employers could not unilaterally declare an impasse, stop bargaining, and raise wages.

Q. 7b If the employers had simply opposed dues check-off without a reason, they may still have been found by the Board to have engaged in bad faith bargaining for opposing a union demand without any reason.

Q. 8a. Negligence or carelessness in bargaining is not a ULP, bad faith bargaining is a ULP. In this case, the Board found that Andrews and negotiator Nassif were negligent in preparing to bargain, and that they evidenced bad faith by their negligence. However, by looking at the total circumstances in the context of a long bargaining history, the Board decided that this incident alone did not indicat bad faith overall.

Q. 8b Having the Board look at contract offers and counteroffers could be argued to be an intrusion into the bargaining process, which the Board is not a part of. However, the Board must see the proposals being offered in order to determine whether there was bad faith bargaining (e.g., was a party advancing unlawful arguments or demands).

Q. 9. If the existing contract included a no strike no lockout clause, then the company could not conduct a lockout during the contract period. The company could conduct a lockout after the contract expires, or if it gave the union notice of its desire to terminate or modify the contract and bargained in good faith for 60 days, it could conduct a lockout after 60 days had passed.

If the contract hasn't expired or 60 days notice hasn't run, then the company can only conduct a defensive lockout. If the employer's motive is to put economic pressure on the union, then the lockout isn't defensive. An employer may conduct a defensive lockdown if its motive is to protect itself economically. If the company had evidence of an imminent strike before the contract or the "cooling off" period expired, then it could defend itself with a lockout.

Since the purpose of the 60 day "cooling off" period is to promote good faith bargaining, an illegal lockout constitutes bad faith bargaining and is a violation of §1153(e).

Q. 10. A labor camp is home for the workers who live there, so it could be argued that the same principles would apply as to residential picketing and that a violation of §1154(a)(1) would occur. However, a labor camp usually houses more than one family and could house hundreds of families. In these circumstances, there could be less expectation of privacy and similar picketing might be less coercive and intimidating. Picketing outside a labor camp would probably be further away and possibly not even audible to the residential targets.

Q. 11a. The Board held that a union's ability to enforce disciplinary rules during a strike is necessary for a union to fulfill its role as collective bargaining representative. The Board then concluded that the legislative history of §1153(c) and the language of §1152 indicate the Legislature's intention that the collective interest of the union and its striking members takes precedence over the right of an individual union member to refrain from engaging in union activity. However, a Federal District Court in 1985 ruled that the ALRA's good standing clause was unconstitutional because it violates First Amendment guarantees of freedom of association. This decision is being appealed.

Q. 11b. No, the ALRB has indicated in dicta that such a rule prohibiting union members from filing charges against the union with the ALRB is not reasonable.

Q. 11c. The ALRB ruled that Martinez did not get a fair trial because he did not understand the procedures concerning representation and giving evidence. The ALRB found the UFW guilty of unlawfully depriving Martinez of his good standing in the UFW, and ordered the UFW to restore him to UFW membership and make him whole for any economic losses he suffered. The ALRB also found that employer Mann Packing had violated ALRA §1153(c) and (a) by discharging Martinez as the UFW requested, and the ALRB ordered Mann Packing to reinstate Martinez.

NOTES

1. Section 1154(b) prohibits a union from engaging in the same conduct.
2. These last two incidents occurred after the alleged unfair labor practices, and so could not have been motivating factors.

3. The table below compares the cost of equipment supplied by Sanchez with its cost when purchased elsewhere:

Item	Sanchez' Price	Price Elsewhere
Gloves	$1.00	$1.19-$1.39
Sleeves	$4.50	$6.00
Clippers	$8.50	$11.00-$11.50
Sacks	$16.00-$18.00	$27.00-$27.50

4. All dates refer to 1981, unless otherwise specified.
5. A no-strike, no-lockout provision in a collective bargaining agreement is a promise that during the duration of the agreement the union will not conduct a strike and the management will not conduct a lockout.]

Table of Cases

Glossary

ALRA —Agricultural Labor Relations Act, Cal. Lab. Code §1140 et seq.

ALRB charge—an allegation filed with the ALRB General Counsel's office that an employer or union has committed an unfair labor practice.

ALRB complaint—a charge that the General Counsel's office has determined has merit and has set for hearing.

ALRB—Agricultural Labor Relations Board.

Authorization card—a card signed by a worker authorizing a union to act on that worker's behalf in collective bargaining. Under the ALRA an election petition must be signed or accompanied by authorization cards signed by 50 percent of the work force.

Bargaining unit—the group of employees with a "community of interest" who vote for a union and are covered by a collective bargaining agreement.

Boycott, consumer—an action by a union to encourage consumers to refrain from purchasing products of the employer in order to induce bargaining concessions from the employer.

Boycott, secondary—exertion of economic pressure by a union on a third party to cease doing business with the employer.

Bracero—literally "arm" in Spanish. A worker under the bracero program, which admitted 4.6 million Mexican workers to the United States between 1942 and 1964 (some returned year after year).

Cease and desist order—an order issued by the ALRB demanding that an employer or union discontinue an activity that is an unfair labor practice.

Certification bar—the ALRA prohibits a representation or decertification election for 12 months following the certification of a union in a bargaining unit representation election held in the same bargaining unit.

Challenged ballot—a ballot voted by a worker who has been challenged either by the employer, union or ALRB representative as not eligible to vote. Challenged ballots are separated from other ballots, and challenges are resolved only if their number would affect the outcome of the election.

Clayton Antitrust Act—prohibits injunctions against unions unless irreparable injury is shown or there is no alternative remedy available.

Closed shop—provides that only union members will be considered for employment. A closed shop agreement is illegal under the NLRA and ALRA.

Collective bargaining—negotiations between an employer and union over wages and working conditions.

Constructive discharge—an employer action forcing an employee to quit, usually by making working conditions more onerous or tedious.

Contract bar—prohibits, a decertification election under the ALRA, when the employer and union have a contract in effect, until the last 12 months of the contract.

Counterproposal—a contract proposal in response to a proposal from the other party to negotiations.

CPD—Citizens Participation Day. A typical United Farmworkers Union contract includes holiday pay for CPD; workers get paid when they do not work for the first Sunday in July, and the employer forwards the pay to the UFW's Martin Luther King charitable fund.

Decertification election—an election in which employees can choose to retain their certified bargaining representative, replace it, or choose not to be represented by a union.

Decision bargaining—bargaining over the decision to subcontract work to nonunion employees.

Dilatory tactics—an indication of bad-faith bargaining where one side delays negotiations in an effort to postpone reaching an agreement.

Due process—the right of union members to a fair hearing and an appeal procedure before expulsion or suspension from a labor union.

Effects bargaining—bargaining over the effects of an employer's decision to discontinue a crop or partially go out of business.

Elasticity of demand for labor—the response of employment to a change in wages.

Election bar—a bar to elections on a farm where a representation election has been conducted within 12 months.

Election petition—a form filed with an ALRB regional office, accompanied by signatures of 50 percent of the workforce, requesting that an representation election be held.

Farm Labor Contractor—a middle-man who contracts with farmers to provide labor for a fee to harvest crops or perform other tasks.

Field packing—the process of sorting and packing crops in the field instead of in packing plants.

Fringe benefit—an employment benefit given in addition to wages. Typical UFW fringe benefits include hourly payments to a health plan and pension plan, vacation pay and paid holidays.

Good-faith bargaining—negotiation with an open mind and a willingness to make reasonable efforts to reach an agreement.

Good standing—valid membership in a labor union. The NLRA prevents a union from revoking good standing except for failure to pay dues and initiation fees. The ALRA allows revocation of membership for failure to follow any reasonable membership requirement if the employee is given due process.

Green carder—an immigrant who has the right to live and work in the United States and, after 5 years, apply to become a naturalized U.S. citizen.

Grievance procedure—the procedures in a collective agreement bargaining agreement to settle disputes arising out of contract interpretation and application.

Hiring Hall—the system of hiring union workers through a central union office.

Hot cargo agreement—an agreement between an employer and union that the employer will not do business with another producer, processor or retailer.

Impasse—a failure to reach agreement between an employer and union despite good faith efforts to do so; an employer may lawfully implement the final proposal after impasse.

Inelastic—little change in employment as wages change.

Injunction—a court order causing an action to occur or preventing an action from taking place.

INS—Immigration and Naturalization Service

Lock-out—an employer attempt to exert bargaining power by laying workers off until an agreement is reached.

Makewhole—a legal remedy for bad-faith bargaining, available only under the ALRA, that attempts to put the employees in the same financial position they would have been in if the employer had bargained in good-faith.

Mandatory subject of bargaining—subjects over which a failure to bargain is an unfair labor practice.

Mechanization—substituting machines for hand labor to produce crops.

Multiple cropping—the practice of growing more than one crop on the same land during a year.

NIO—Notice of Intent to Organize

NLRA—National Labor Relations Act

NLRB—National Labor Relations Board

NOA—Notice of Intent to Obtain Access

Norris La Guardia Act—a Congressional Act that reduced the power of courts to issue injunctions in labor disputes.

Notice to workers—an ALRB remedy that informs workers of the outcome of an unfair labor practice complaint and hearing. It is typically read aloud to workers and mailed to worker's homes.

Peak farm employment—the maximum number of employees employed at one time during the year.

Peak-trough ratio—the ratio of peak employment to the minimum number of workers employed during one year.

Permanent replacement worker—a worker replacing a striking worker when it is necessary for the employer to guarantee permanent employment in order to obtain workers. Under the ALRA strikers can "bump" permanent replacement workers the next season after the strike.

Permissive subjects of bargaining—subjects which either party can refuse to bargain over without committing a ULP.

Picketing—the patrolling of a struck farm by union members or sympathizers to protest a ULP, advertise a strike, or encourage strikebreakers to join the strike.

Picketing, information—picketing to inform the public of a labor dispute. It may occur off of the farm.

Picketing, recognitional—picketing by a union to force an employer to recognize it as the worker's exclusive bargaining representative. Since a union can only become a certified bargaining representative under the ALRA by winning an election, recognitional picketing is prohibited.

Picketing, residential—picketing of a worker's residences, usually to encourage them to join a strike.

Pre-election conference—a conference between an ALRB representative, the employer representative, and the union representative held before a representative or decertification election in order to discuss election details.

Precedent—a judicial or administrative decision that may be used as a standard in subsequent similar cases.

Preconditions—an event that must occur before one party will begin to bargain. Attaching preconditions to an offer to bargain is evidence of bad-faith bargaining.

Prohibited subjects of bargaining—subjects which, even if agreed to, will not be enforced by the courts.

Protected concerted activity—an activity, usually by two or more employees, to protest terms, tenure, or conditions of employment.

Quality circle—a committee of workers and management personnel that cooperates in certain decision making functions.

Reinstatement with back pay—an ALRB remedy for unlawful discrimination in conditions of employment (Sec 1153(c) violation). The ALRB orders the employer to rehire the discriminatee and to pay that person an amount equal to what the employee lost as a result of the employer's unlawful action.

Remedy—an ALRB order designed to restore the status quo after an unfair labor practice.

Representation election—an election conducted by the ALRB or NLRB to determine if employees desire to be represented by a collective bargaining representative, and if so, which union should be the representative.

Runaway shop—a farm that transfers its work elsewhere in order to avoid unionization.

Strike—a concerted stopping of work in order to compel an employer to agree to workers' demands.

Strike, economic—a strike to pressure the employer to agree to worker or union demands over wages or working conditions.

Strike, intermittent—an unlawful series of work stoppages.

Strike, jurisdictional—a strike caused by a dispute between two unions over which union's members are entitled to certain work assignment.

Strike, partial—a work slowdown or partial work stoppage.

Strike, primary—strike over a dispute between a primary employer and its employees.

Strike, recognitional—a strike to force the employer to recognize a union as the exclusive bargaining representative of the workers. Under the ALRA this type of strike is illegal.

Strike, secondary—a strike to force the employer to stop doing business with another company, the primary employer, with whom the union has a labor dispute. Secondary strikes are prohibited by the ALRA and NLRA.

Strike, unfair labor practice—a strike in response to an unfair labor practice committed by the employer.

Strikebreaking—working at a farm while a strike is in progress.

Surface bargaining—negotiating with a union or employer while attempting to prevent the negotiations from reaching an agreement. This is sometimes done by employers who wish to delay wage increases resulting from the union contract.

Sweet-heart contract—a contract negotiated by union leaders which provides benefits to union leaders at the expense of union members benefits.

Taft-Hartley Act—a Congressional Act that amended the National Labor Relations Act in 1947 to establish union unfair labor practices and decertification elections.

Technical objection—a claim that an election was not valid and should not be certified because the employer specified was incorrect, employment was less than 50 percent of peak when the election was held, the bargaining unit was incorrect, or ineligible workers voted.

Technical refusal to bargain—an employer's refusal to bargain with a certified union because the employer believes the ALRB erred in certifying the union. When the union files a charge of bad-faith bargaining against the employer, the employer raises the technical refusal to bargain defense at the appeal court level—the employer asserts that the union should not have been certified and that therefore the employer has no duty to bargain.

Temporary replacement worker—under the NLRA a striker is entitled to his job back after unconditionally offering to return to work if he has been replaced by a temporary replacement worker. Under the ALRA replacement workers are presumed to be permanent for that season, but strikers are entitled to reinstatement the next season.

Unfair labor practice—employer and union actions which violate the rights of workers created by the ALRA and NLRA.

Unilateral changes in wage or working conditions—changes in wages or working conditions by the employer without the permission of the certified bargaining agent. The ALRA permits unilateral changes only after a bonafide impasse in bargaining, and only up to the amount of the employer's last proposal.

Union—an organization of workers formed to promote their collective interests with respect to wages and working conditions.

Union security clause—part of a collective bargaining agreement under which the employer helps the union to maintain membership by deducting or checking-off union dues. The clause requires the employer to discharge workers who are not in good standing with the union.

Wagner Act—also known as the National Labor Relations Act, it established the National Labor Relations Board in 1935.

Index

Printed and bound by CPI Group (UK) Ltd, Croydon, CR0 4YY

23/10/2024

01778241-0012